HUMANITIES SOCIETY AND TECHNOLOGY

LIVING WITH CHANGE

REVISED PRINTING

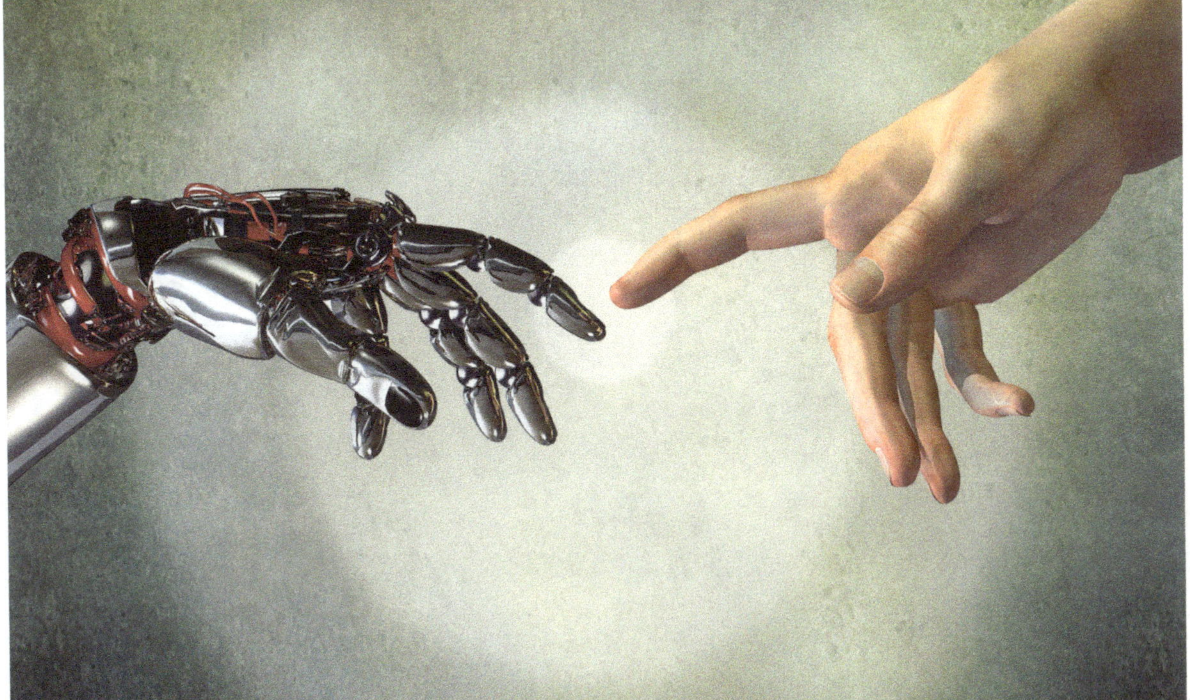

Robin Satterwhite | Duane Galloway | Nettie Amirtharaj

Rowan Cabarrus Community College

Kendall Hunt
publishing company

Kendall Hunt
publishing company

www.kendallhunt.com
Send all inquiries to:
4050 Westmark Drive
Dubuque, IA 52004-1840

Revised Printing: 2016

ISBN 978-1-4652-9661-0

Printed in the United States of America

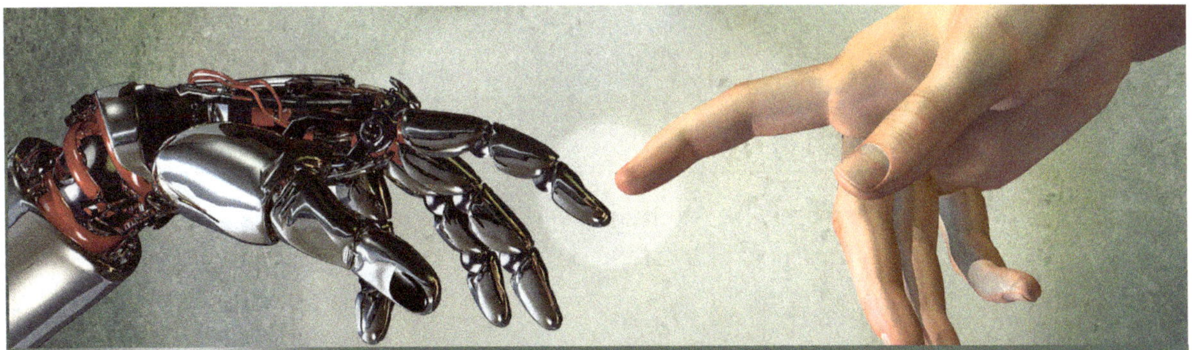

Contents

Introduction

Greetings Humanities Students! Technology and Society, HUM 110, has been offered at Rowan Cabarrus Community College since 2001. Many of the college's academic programs encourage or require students to enroll in the course as a required Humanities elective.

Course instructors have struggled since 2001 to find appropriate textbooks for Humanities 110. Although many are available, no specific textbook seemed to quite fit the course description. Additionally, each Technology and Society instructor taught the course's objectives with their own unique style; some approaching the course from a historical standpoint, a few from a literary perspective, and others as somewhat of a combination. The subject of how humans have reacted and responded to technology through the years is broad, yet it can be approached and studied through many different lenses.

While using various texts, our institution's instructors began to bring in or introduce their own materials in addition to the adopted textbook. Some instructors used essays as supplemental assignments while others distributed fictional writings and websites for review. These supplements were all related to course objectives, while concentrating on how society reacts to technological advances.

Therefore the idea for a customized anthology was born. The authors hope students find this text informative and thought provoking. The text remains a work in progress, and one that welcomes feedback from its readers. It is also the writers' intent that this book will challenge students to critically analyze technology and how humans live with and react to it.

Sincerely,
Robin G. Satterwhite
Humanities Department Chair
Rowan Cabarrus Community College

SECTION 1 What Are Humanities? What Is Technology? What Is Society?

© Angela Waye/Shutterstock.com

Introduction

Humanities, technology, and society are terms that are used often, but are difficult terms to define. The Humanities is a relatively broad subject that can be described as an emotional release that helps people express the seemingly inexpressible. Oftentimes these feelings are voiced through literature, art, music, drama, philosophy, and various offshoots of these mediums. For example, the Industrial Revolution was a time of tremendous change because the innovative machines that were being created were suddenly transforming the lives of all classes of people. The Humanities allowed the people caught in this tumultuous period in history to express their fears and frustrations about the future.

Similarly, technology is a term that is frequently misinterpreted and misunderstood. Technology is anything that helps a group complete a task more easily. In modern times, many view technology as only electronic devices such as cell phones, computers, and televisions, but all technologies cannot be grouped into electronic technologies. For example, a pencil is a technology because it helps people communicate through writing. Similarly, a rock is a technology if it is used as a weapon, a tool, or a material to build a house. As you will see through the readings in this book, there have been a multitude of technological achievements over time, some seemingly beneficial and others problematic; however, controversy over the impact of each technology in history exists depending on each person's experience with it.

In this book, we will discuss how societies are impacted by such technologies. The term society refers to a structured system of organization. The members of a society typically share some type of territory, interact with each other, and share many of the same dominant beliefs. Oftentimes, they come together with a single purpose in mind, survival. When they attempt to procure ways to ensure survival and increase productivity, society and technology meet.

WHEN I HEARD THE LEARN'D ASTRONOMER

Walt Whitman

When I heard the learn'd astronomer,
When the proofs, the figures, were ranged in columns before me,
When I was shown the charts and diagrams, to add, divide, and measure them,
When I sitting heard the astronomer where he lectured with much applause in the
    lecture-room,
How soon unaccountable I became tired and sick,
Till rising and gliding out I wander'd off by myself,
In the mystical moist night-air, and from time to time,
Look'd up in perfect silence at the stars.

Name ________________________ Course ______________ Date ________________

Chapter Questions

1. In this story, what are Whitman's feelings about the formulaic approach of the astronomer?

SECTION 2 The Industrial Revolution

© Everett Historical/Shutterstock.com

One would be hard-pressed to find a phenomenon that has impacted the human experience quite like the Industrial Revolution. The scientific and technological innovations that allowed the conversion from separate, handicraft operations to a concentrated, streamlined, mechanical process touched all classes. The Humanities allowed people caught in this whirlwind of rapid change to express their fears and frustrations.

The Industrial Revolution Changes Work

The work experience is one area where the frustration of change was most keenly felt, and it is around work that most of the selections in this chapter revolve. Not long after the Industrial Revolution began in Britain in the early 1700s, scholars began to identify in factory workers an emotional disconnect from work which they labeled "alienation." Craftsmen and farmers, who made up the majority of the pre-industrial workforce, had usually witnessed the entire production process, whether they shaped a chair from a tree trunk or coaxed a fragile seed into a shaft of wheat. This led to a sense of pride, an emotional and visceral connection to the product. However, compartmented on an assembly line, the industrial worker saw only one part of the production process—that process now earning much more of their attention than the product. The Industrial worker likewise lost a sense of immediate autonomy over the pace and start/stop time of work. While their craftsmen/farmer forefathers ordered their days by the seasons and the sun the Industrial worker was forced to learn the "telling" of abstract, measured time. Like the workers' very movements, time had become a commodity to be bought and sold. As the Industrial Revolution spread from Britain to the United States and the other west-European countries, workers also experienced dangerous conditions and pay unequal to the wealth generated by their labor, especially the women and children who seemed to make up a growing percentage of the workforce.

Societal impact

Despite this, thousands of families migrated from the countryside to newly burgeoning cities where many of the factories were located. While craftsmen whose skills had been replaced by automation came by necessity, others were attracted to the steadier pay of the factories compared to farming. The horde of job-seekers stretched already thin city infrastructures. Slums sprang up, improper water and sewer systems killed thousands each year, and fire and police protection lagged as disasters like New York's Triangle Shirtwaist Factory fire in 1911 demonstrate.

Today, these "growing pains" are evident in developing countries where the Industrial Revolution began later. The collapse of the Rana Plaza garment factory in Bangladesh in April of 2013, which killed over 1,100 workers, was eerily familiar to the Western countries who now haltingly navigate a "Post-Industrial Age." The Industrial Revolution did dramatically raise the overall wealth of the nations where it started. Innovations in machines, plastics, medicines, and electronics made them cheaper and safer and, over time, helped the average life span to double in industrialized nations. However, the alienation, stress, "tyranny of time," and continued pollution of air and water still plague the post-Industrial world. Both wealthy and developing nations feel the vicissitudes of a world economy that seems, at least on the level of the workers, to be unpredictable and uncontrollable. The Humanities continue to ask if these problems are worth the rewards. Does a paycheck balance a worker's loss of autonomy and identity? Is an overall increase in wealth mitigated by a polluted environment and hazardous work conditions? In this vital role, the Humanities always question our answers more than they answer our questions.

HARD TIMES

Chapter II: Murdering the Innocents

Charles Dickens

After his father was incarcerated in a debtor's prison in 1824, Charles Dickens left school to work in a factory. In most of his novels, Dickens describes the working conditions in factories and the discrepancy between the upper and lower classes during the Industrial Revolution that he saw first-hand. His novel, Hard Times, vividly describes the social condition of poverty during the Industrial Revolution. Early in the novel, Thomas Gradgrind, a wealthy merchant in the grim, oppressive industrial town of Coketown, and one of the stories' main characters, establishes a school. In this school, students are taught facts and figures rather than engaging in whimsical and artistic thoughts. The fourth and fifth chapter of the novel describe Gradgrind speaking to and quizzing some of his pupils.

THOMAS GRADGRIND, sir. A man of realities. A man of facts and calculations. A man who proceeds upon the principle that two and two are four, and nothing over, and who is not to be talked into allowing for anything over. Thomas Gradgrind, sir—peremptorily Thomas—Thomas Gradgrind. With a rule and a pair of scales, and the multiplication table always in his pocket, sir, ready to weigh and measure any parcel of human nature, and tell you exactly what it comes to. It is a mere question of figures, a case of simple arithmetic. You might hope to get some other nonsensical belief into the head of George Gradgrind, or Augustus Gradgrind, or John Gradgrind, or Joseph Gradgrind (all supposititious, non-existent persons), but into the head of Thomas Gradgrind—no, sir!

In such terms Mr. Gradgrind always mentally introduced himself, whether to his private circle of acquaintance, or to the public in general. In such terms, no doubt, substituting the words "boys and girls," for "sir," Thomas Gradgrind now presented Thomas Gradgrind to the little pitchers before him, who were to be filled so full of facts.

Indeed, as he eagerly sparkled at them from the cellarage before mentioned, he seemed a kind of cannon loaded to the muzzle with facts, and prepared to blow them clean out of the regions of childhood at one discharge. He seemed a galvanizing apparatus, too, charged with a grim mechanical substitute for the tender young imaginations that were to be stormed away.

"Girl number twenty," said Mr. Gradgrind, squarely pointing with his square forefinger, "I don't know that girl. Who is that girl?"

"Sissy Jupe, sir," explained number twenty, blushing, standing up, and curtseying.

"Sissy is not a name," said Mr. Gradgrind. "Don't call yourself Sissy. Call yourself Cecilia."

"It's father who calls me Sissy, sir," returned the young girl in a trembling voice, and with another curtsey.

"Then he has no business to do it," said Mr. Gradgrind. "Tell him he mustn't. Cecilia Jupe. Let me see. What is your father?"

"He belongs to the horse-riding, if you please, sir."

Mr. Gradgrind frowned, and waved off the objectionable calling with his hand.

"We don't want to know anything about that, here. You mustn't tell us about that, here. Your father breaks horses, doesn't he?"

"If you please, sir, when they can get any to break, they do break horses in the ring, sir."

"You mustn't tell us about the ring, here. Very well, then. Describe your father as a horsebreaker. He doctors sick horses, I dare say?"

"Oh yes, sir."

"Very well, then. He is a veterinary surgeon, a farrier, and horsebreaker. Give me your definition of a horse."

(Sissy Jupe thrown into the greatest alarm by this demand.)

"Girl number twenty unable to define a horse!" said Mr. Gradgrind, for the general behoof of all the little pitchers. "Girl number twenty possessed of no facts, in reference to one of the commonest of animals! Some boy's definition of a horse. Bitzer, yours."

© Kevin Eaves/Shutterstock.com

The square finger, moving here and there, lighted suddenly on Bitzer, perhaps because he chanced to sit in the same ray of sunlight which, darting in at one of the bare windows of the intensely white-washed room, irradiated Sissy. For, the boys and girls sat on the face of the inclined plane in two compact bodies, divided up the center by a narrow interval; and Sissy, being at the corner of a row on the sunny side, came in for the beginning of a sunbeam, of which Bitzer, being at the corner of a row on the other side, a few rows in advance, caught the end. But, whereas the girl was so dark-eyed and dark-haired, that she seemed to receive a deeper and more lustrous color from the sun, when it shone upon her, the boy was so light-eyed and light-haired that the self-same rays appeared to draw out of him what little color he ever possessed. His cold eyes would hardly have been eyes, but for the short ends of lashes which, by bringing them into immediate contrast with something paler than themselves, expressed their form. His short-cropped hair might have been a mere continuation of the sandy freckles on his forehead and face. His skin was so unwholesomely deficient in the natural tinge that he looked as though, if he were cut, he would bleed white.

"Bitzer," said Thomas Gradgrind. "Your definition of a horse."

"Quadruped. Graminivorous. Forty teeth, namely twenty-four grinders, four eye-teeth, and twelve incisive. Sheds coat in the spring; in marshy countries, sheds hoofs, too. Hoofs hard, but requiring to be shod with iron. Age known by marks in mouth." Thus (and much more) Bitzer.

"Now girl number twenty," said Mr. Gradgrind. "You know what a horse is."

She curtseyed again, and would have blushed deeper, if she could have blushed deeper than she had blushed all this time. Bitzer, after rapidly blinking at Thomas Gradgrind with both eyes at once, and so catching the light upon his quivering ends of lashes that they looked like the antennæ of busy insects, put his knuckles to his freckled forehead, and sat down again.

The third gentleman now stepped forth. A mighty man at cutting and drying, he was; a government officer; in his way (and in most other people's too), a professed pugilist; always in training, always with a system to force down the general throat like a bolus, always to be heard of at the bar of his little Public-office, ready to fight all England. To continue in fistic phraseology, he had a genius for coming up to the scratch, wherever and whatever it was, and proving himself an ugly customer. He would go in and damage any subject whatever with his right, follow up with his left, stop, exchange, counter, bore his opponent (he always fought All England) to the ropes, and fall upon him neatly. He was certain to knock the wind out of common sense, and render that unlucky adversary deaf to the call of time. And he had it in charge from high authority to bring about the great public-office Millennium, when Commissioners should reign upon earth.

"Very well," said this gentleman, briskly smiling, and folding his arms. "That's a horse. Now, let me ask you girls and boys, Would you paper a room with representations of horses?"

After a pause, one half of the children cried in chorus, "Yes, sir!" Upon which the other half, seeing in the gentleman's face that Yes was wrong, cried out in chorus, "No, sir!"—as the custom is, in these examinations.

"Of course, No. Why wouldn't you?"

A pause. One corpulent slow boy, with a wheezy manner of breathing, ventured the answer, Because he wouldn't paper a room at all, but would paint it.

"You *must* paper it," said the gentleman, rather warmly.

"You must paper it," said Thomas Gradgrind, "whether you like it or not. Don't tell *us* you wouldn't paper it. What do you mean, boy?"

"I'll explain to you, then," said the gentleman, after another and a dismal pause, "why you wouldn't paper a room with representations of horses. Do you ever see horses walking up and down the sides of rooms in reality—in fact? Do you?"

"Yes, sir!" from one half. "No, sir!" from the other.

"Of course no," said the gentleman, with an indignant look at the wrong half. "Why, then, you are not to see anywhere, what you don't see in fact; you are not to have anywhere, what you don't have in fact. What is called Taste, is only another name for Fact." Thomas Gradgrind nodded his approbation.

"This is a new principle, a discovery, a great discovery," said the gentleman. "Now, I'll try you again. Suppose you were going to carpet a room. Would you use a carpet having a representation of flowers upon it?"

There being a general conviction by this time that "No, sir!" was always the right answer to this gentleman, the chorus of No was very strong. Only a few feeble stragglers said Yes: among them Sissy Jupe.

"Girl number twenty," said the gentleman, smiling in the calm strength of knowledge.

Sissy blushed, and stood up.

"So you would carpet your room—or your husband's room, if you were a grown woman, and had a husband—with representations of flowers, would you?" said the gentleman. "Why would you?"

"If you please, sir, I am very fond of flowers," returned the girl.

"And is that why you would put tables and chairs upon them, and have people walking over them with heavy boots?"

"It wouldn't hurt them, sir. They wouldn't crush and wither, if you please, sir. They would be the pictures of what was very pretty and pleasant, and I would fancy—"

"Ay, ay, ay! But you mustn't fancy," cried the gentleman, quite elated by coming so happily to his point. "That's it! You are never to fancy."

"You are not, Cecilia Jupe," Thomas Gradgrind solemnly repeated, "to do anything of that kind."

"Fact, fact, fact!" said the gentleman. And "Fact, fact, fact!" repeated Thomas Gradgrind.

"You are to be in all things regulated and governed," said the gentleman, by fact. We hope to have, before long, a board of fact, composed of commissioners of fact, who will force the people to be a people of fact, and of nothing but fact. You must discard the word Fancy altogether. You have nothing to do with it. You are not to have, in any object of use or ornament, what would be a contradiction in fact. You don't walk upon flowers in fact; you cannot be allowed to walk upon flowers in carpets. You don't find that foreign birds and butterflies come and perch upon your crockery; you cannot be permitted to paint foreign birds and butterflies upon your crockery. You never meet with quadrupeds going up and down walls; you must not have quadrupeds represented upon walls. You must use," said the gentleman, "for all these purposes, combinations and modifications (in primary colors) of mathematical figures that are susceptible of proof and demonstration. This is the new discovery. This is fact. This is taste."

The girl curtseyed, and sat down. She was very young, and she looked as if she were frightened by the matter-of-fact prospect the world afforded.

"Now, if Mr. M'Choakumchild," said the gentleman, "will proceed to give his first lesson here, Mr. Gradgrind, I shall be happy, at your request, to observe his mode of procedure."

Mr. Gradgrind was much obliged. "Mr. M'Choakumchild, we only wait for you."

So, Mr. M'Choakumchild began in his best manner. He and some one hundred and forty other schoolmasters, had been lately turned at the same time, in the same factory, on the same principles, like so many pianoforte legs. He had been put through an immense variety of paces, and had answered volumes of head-breaking questions. Orthography, etymology, syntax, and prosody, biography, astronomy, geography, and general cosmography, the sciences of compound proportion, algebra, land-surveying and leveling, vocal music, and drawing from models, were all at the ends of his ten chilled fingers. He had worked his stony way into Her Majesty's most Honorable Privy Council's Schedule B, and had taken the bloom off the higher branches of mathematics and physical science, French, German, Latin, and Greek. He knew all about all the Water Sheds of the entire world (whatever they are), and all the histories of all the peoples, and all the names of all the rivers and mountains, and all the productions, manners, and customs of all the countries, and all their boundaries and bearings on the two and thirty points of the compass. Ah, rather overdone, M'Choakumchild. If he had only learnt a little less, how infinitely better he might have taught much more!

He went to work in this preparatory lesson, not unlike Morgiana in the Forty Thieves: looking into all the vessels ranged before him, one after another, to see what they contained. Say, good M'Choakumchild. When from thy boiling store, thou shalt fill each jar brim full by-and-by, dost thou think that thou wilt always kill outright the robber Fancy lurking within—or sometimes only maim him and distort him!

Name ____________________ Course ____________ Date ______________

Chapter Questions

1. How does this chapter from the book, which is set in the school, indicate Gradgrind's philosophy on learning? Give specific examples mentioned in the reading.

2. How does the relationship between Gradgrind and the children mirror the relationship between merchants and the working class during the Industrial Revolution?

THE APOSTATE

Jack London

"The *Apostate*" tells the story of Johnny, a young boy who has been forced to work in a factory during the early 20th century to support his family. Because he has been working in a dangerous, strenuous job since he was seven years old, his childhood has been almost nonexistent. The textile factories, where Johnny works in this story, were particularly unsafe for all factory workers during Industrial Revolution. Many factory owners used children as laborers because of their small size. They were more easily able to do tasks like crawling under machines to fix problems, and owners did not have to pay them as much as adults, even if they were doing a similar job. The author of the piece, Jack London, frequently acted as a social activist during this time period fighting for workers' rights. In this story, he provides realistic and detailed portrayals of the working conditions for the poor to enact reform as a social activist and attempts to call people's attention to the horrific working and living conditions that were forced upon them.

"Now I wake me up to work;
I pray the Lord I may not shirk.
If I should die before the night,
I pray the Lord my work's all right.
Amen."

"If you don't git up, Johnny, I won't give you a bite to eat!"

The threat had no effect on the boy. He clung stubbornly to sleep, fighting for its oblivion as the dreamer fights for his dream. The boy's hands loosely clenched themselves, and he made feeble, spasmodic blows at the air. These blows were intended for his mother, but she betrayed practised familiarity in avoiding them as she shook him roughly by the shoulder.

"Lemme 'lone!"

It was a cry that began, muffled, in the deeps of sleep, that swiftly rushed upward, like a wail, into passionate belligerence, and that died away and sank down into an inarticulate whine. It was a bestial cry, as of a soul in torment, filled with infinite protest and pain.

But she did not mind. She was a sad-eyed, tired-faced woman, and she had grown used to this task, which she repeated every day of her life. She got a grip on the bedclothes and tried to strip them down; but the boy, ceasing his punching, clung to them desperately. In a huddle, at the foot of the bed, he still remained covered. Then she tried dragging the bedding to the floor. The boy opposed her. She braced herself. Hers was the superior weight, and the boy and bedding gave, the former instinctively following the latter in order to shelter against the chill of the room that bit into his body.

As he toppled on the edge of the bed it seemed that he must fall head-first to the floor. But consciousness fluttered up in him. He righted himself and for a moment perilously balanced. Then he struck the floor on his feet. On the instant his mother seized him by the shoulders and shook him. Again his fists struck out, this time with more force and directness. At the same time his eyes opened. She released him. He was awake.

"All right," he mumbled.

She caught up the lamp and hurried out, leaving him in darkness.

"You'll be docked," she warned back to him.

He did not mind the darkness. When he had got into his clothes, he went out into the kitchen. His tread was very heavy for so thin and light a boy. His legs dragged with their own weight, which seemed unreasonable because they were such skinny legs. He drew a broken-bottomed chair to the table.

"Johnny," his mother called sharply.

He arose as sharply from the chair, and, without a word, went to the sink. It was a greasy, filthy sink. A smell came up from the outlet. He took no notice of it. That a sink should smell was to him part of the natural order, just as it was a part of the natural order that the soap should be grimy with dish-water and hard to lather. Nor did he try very hard to make it lather. Several splashes of the cold water from the running faucet completed the function. He did not wash his teeth. For that matter he had never seen a toothbrush, nor did he know that there existed beings in the world who were guilty of so great a foolishness as tooth washing.

"You might wash yourself wunst a day without bein' told," his mother complained.

She was holding a broken lid on the pot as she poured two cups of coffee. He made no remark, for this was a standing quarrel between them, and the one thing upon which his mother was hard as adamant. "Wunst" a day it was compulsory that he should wash his face. He dried himself on a greasy towel, damp and dirty and ragged, that left his face covered with shreds of lint.

"I wish we didn't live so far away," she said, as he sat down. "I try to do the best I can. You know that. But a dollar on the rent is such a savin', an' we've more room here. You know that."

He scarcely followed her. He had heard it all before, many times. The range of her thought was limited, and she was ever harking back to the hardship worked upon them by living so far from the mills.

"A dollar means more grub," he remarked sententiously. "I'd sooner do the walkin' an' git the grub."

He ate hurriedly, half chewing the bread and washing the unmasticated chunks down with coffee. The hot and muddy liquid went by the name of coffee. Johnny thought it was coffee—and excellent coffee. That was one of the few of life's illusions that remained to him. He had never drunk real coffee in his life.

In addition to the bread, there was a small piece of cold pork. His mother refilled his cup with coffee. As he was finishing the bread, he began to watch if more was forthcoming. She intercepted his questioning glance.

"Now, don't be hoggish, Johnny," was her comment. "You've had your share. Your brothers an' sisters are smaller'n you."

He did not answer the rebuke. He was not much of a talker. Also, he ceased his hungry glancing for more. He was uncomplaining, with a patience that was as terrible as the school in which it had been learned. He finished his coffee, wiped his mouth on the back of his hand, and started to rise.

"Wait a second," she said hastily. "I guess the loaf kin stand you another slice—a thin un."

There was legerdemain in her actions. With all the seeming of cutting a slice from the loaf for him, she put loaf and slice back in the bread box and conveyed to him one of her own two slices. She believed she had deceived him, but he had noted her sleight-of-hand. Nevertheless, he took the bread shamelessly. He had a philosophy that his mother, because of her chronic sickliness, was not much of an eater anyway.

She saw that he was chewing the bread dry, and reached over and emptied her coffee cup into his.

"Don't set good somehow on my stomach this morning," she explained.

A distant whistle, prolonged and shrieking, brought both of them to their feet. She glanced at the tin alarm-clock on the shelf. The hands stood at half-past five. The rest of the factory world was just arousing from sleep. She drew a shawl about her shoulders, and on her head put a dingy hat, shapeless and ancient.

"We've got to run," she said, turning the wick of the lamp and blowing down the chimney.

They groped their way out and down the stairs. It was clear and cold, and Johnny shivered at the first contact with the outside air. The stars had not yet begun to pale in the sky, and the city lay in blackness. Both Johnny and his mother shuffled their feet as they walked. There was no ambition in the leg muscles to swing the feet clear of the ground.

After fifteen silent minutes, his mother turned off to the right.

"Don't be late," was her final warning from out of the dark that was swallowing her up.

He made no response, steadily keeping on his way. In the factory quarter, doors were opening everywhere, and he was soon one of a multitude that pressed onward through the dark. As he entered the factory gate the whistle blew again. He glanced at the east. Across a ragged sky-line of housetops a pale light was beginning to creep. This much he saw of the day as he turned his back upon it and joined his work gang.

© Everett Historical/Shutterstock.com

He took his place in one of many long rows of machines. Before him, above a bin filled with small bobbins, were large bobbins revolving rapidly. Upon these he wound the jute-twine of the small bobbins. The work was simple. All that was required was celerity. The small bobbins were emptied so rapidly, and there were so many large bobbins that did the emptying, that there were no idle moments.

He worked mechanically. When a small bobbin ran out, he used his left hand for a brake, stopping the large bobbin and at the same time, with thumb and forefinger, catching the flying end of twine. Also, at the same time, with his right hand, he caught up the loose twine-end of a small bobbin. These various acts with both hands were performed simultaneously and swiftly. Then there would come a flash of his hands as he looped the weaver's knot and released the bobbin. There was nothing difficult about weaver's knots. He once boasted he could tie them in his sleep. And for that matter, he sometimes did, toiling centuries long in a single night at tying an endless succession of weaver's knots.

Some of the boys shirked, wasting time and machinery by not replacing the small bobbins when they ran out. And there was an overseer to prevent this. He caught Johnny's neighbour at the trick, and boxed his ears.

"Look at Johnny there—why ain't you like him?" the overseer wrathfully demanded.

Johnny's bobbins were running full blast, but he did not thrill at the indirect praise. There had been a time. . . but that was long ago, very long ago. His apathetic face was expressionless as he listened to himself being held up as a shining example. He was the perfect worker. He knew that. He had been told so, often. It was a commonplace, and besides it didn't seem to mean anything to him any more. From the perfect worker he had evolved into the perfect machine. When his work went wrong, it was with him as with the machine, due to faulty material. It would have been as possible for a perfect nail-die to cut imperfect nails as for him to make a mistake.

And small wonder. There had never been a time when he had not been in intimate relationship with machines. Machinery had almost been bred into him, and at any rate he had been brought up on it. Twelve years before, there had been a small flutter of excitement in the loom room of this very mill. Johnny's mother had fainted. They stretched her out on the floor in the midst of the shrieking machines. A couple of elderly women were called from their looms. The foreman assisted. And in a few minutes there was one more soul in the loom room than had entered by the doors. It was Johnny, born with the pounding, crashing roar of the looms in his ears, drawing with his first breath the warm, moist air that was thick with flying lint. He had coughed that first day in order to rid his lungs of the lint; and for the same reason he had coughed ever since.

The boy alongside of Johnny whimpered and sniffed. The boy's face was convulsed with hatred for the overseer who kept a threatening eye on him from a distance; but every bobbin was running full. The boy yelled terrible oaths into the whirling bobbins before him; but the sound did not carry half a dozen feet, the roaring of the room holding it in and containing it like a wall.

Of all this Johnny took no notice. He had a way of accepting things. Besides, things grow monotonous by repetition, and this particular happening he had witnessed many times. It seemed to him as useless to oppose the overseer as to defy the will of a machine. Machines were made to go in certain ways and to perform certain tasks. It was the same with the overseer.

But at eleven o'clock there was excitement in the room. In an apparently occult way the excitement instantly permeated everywhere. The one-legged boy who worked on the other side of Johnny bobbed swiftly across the floor to a bin truck that stood empty. Into this he dived out of sight, crutch and all. The superintendent of the mill was coming along, accompanied by a young man. He was well dressed and wore a starched shirt—a gentleman, in Johnny's classification of men, and also, "the Inspector."

He looked sharply at the boys as he passed along. Sometimes he stopped and asked questions. When he did so, he was compelled to shout at the top of his lungs, at which moments his face was ludicrously contorted with the strain of making himself heard. His quick eye noted the empty machine alongside of Johnny's, but he said nothing. Johnny also caught his eye, and he stopped abruptly. He caught Johnny by the arm to draw him back a step from the machine; but with an exclamation of surprise he released the arm.

"Pretty skinny," the superintendent laughed anxiously.

"Pipe stems," was the answer. "Look at those legs. The boy's got the rickets—incipient, but he's got them. If epilepsy doesn't get him in the end, it will be because tuberculosis gets him first."

Johnny listened, but did not understand. Furthermore he was not interested in future ills. There was an immediate and more serious ill that threatened him in the form of the inspector.

"Now, my boy, I want you to tell me the truth," the inspector said, or shouted, bending close to the boy's ear to make him hear. "How old are you?"

"Fourteen," Johnny lied, and he lied with the full force of his lungs. So loudly did he lie that it started him off in a dry, hacking cough that lifted the lint which had been settling in his lungs all morning.

"Looks sixteen at least," said the superintendent.

"Or sixty," snapped the inspector.

"He's always looked that way."

"How long?" asked the inspector, quickly.

"For years. Never gets a bit older."

"Or younger, I dare say. I suppose he's worked here all those years?"

"Off and on—but that was before the new law was passed," the superintendent hastened to add.

"Machine idle?" the inspector asked, pointing at the unoccupied machine beside Johnny's, in which the part-filled bobbins were flying like mad.

"Looks that way." The superintendent motioned the overseer to him and shouted in his ear and pointed at the machine. "Machine's idle," he reported back to the inspector.

They passed on, and Johnny returned to his work, relieved in that the ill had been averted. But the one-legged boy was not so fortunate. The sharp-eyed inspector haled him out at arms length from the bin truck. His lips were quivering, and his face had all the expression of one upon whom was fallen profound and irremediable disaster. The overseer looked astounded, as though for the first time he had laid eyes on the boy, while the superintendent's face expressed shock and displeasure.

"I know him," the inspector said. "He's twelve years old. I've had him discharged from three factories inside the year. This makes the fourth."

He turned to the one-legged boy. "You promised me, word and honour, that you'd go to school."

The one-legged boy burst into tears. "Please, Mr. Inspector, two babies died on us, and we're awful poor."

"What makes you cough that way?" the inspector demanded, as though charging him with crime.

And as in denial of guilt, the one-legged boy replied: "It ain't nothin'. I jes' caught a cold last week, Mr. Inspector, that's all."

In the end the one-legged boy went out of the room with the inspector, the latter accompanied by the anxious and protesting superintendent. After that monotony settled down again. The long morning and the longer afternoon wore away and the whistle blew for quitting time. Darkness had already fallen when Johnny passed out through the factory gate. In the interval the sun had made a golden ladder of the sky, flooded the world with its gracious warmth, and dropped down and disappeared in the west behind a ragged sky-line of housetops.

Supper was the family meal of the day—the one meal at which Johnny encountered his younger brothers and sisters. It partook of the nature of an encounter, to him, for he was very old, while they were distressingly young. He had no patience with their excessive and amazing juvenility. He did not understand it. His own childhood was too far behind him. He was like an old and irritable man, annoyed by the turbulence of their young spirits that was to him arrant silliness. He glowered silently over his food, finding compensation in the thought that they would soon have to go to work. That would take the edge off of them and make them sedate and dignified—like him. Thus it was, after the fashion of the human, that Johnny made of himself a yardstick with which to measure the universe.

During the meal, his mother explained in various ways and with infinite repetition that she was trying to do the best she could; so that it was with relief, the scant meal ended, that Johnny shoved back his chair and arose. He debated for a moment between bed and the front door, and finally went out the latter. He did not go far. He sat down on the stoop, his knees drawn up and his narrow shoulders drooping forward, his elbows on his knees and the palms of his hands supporting his chin.

As he sat there, he did no thinking. He was just resting. So far as his mind was concerned, it was asleep. His brothers and sisters came out, and with other children played noisily about him. An electric globe at the corner lighted their frolics. He was peevish and irritable, that they knew; but the spirit of adventure lured them into teasing him. They joined hands before him, and, keeping time with their bodies, chanted in his face weird and uncomplimentary doggerel. At first he snarled curses at them—curses he had learned from the lips of various foremen. Finding this futile, and remembering his dignity, he relapsed into dogged silence.

His brother Will, next to him in age, having just passed his tenth birthday, was the ringleader. Johnny did not possess particularly kindly feelings toward him. His life had early been embittered by continual giving over and giving way to Will. He had a definite feeling that Will was greatly in his debt and was ungrateful about it. In his own playtime, far back in the dim past, he had been robbed of a large part of that playtime by being compelled to take care of Will. Will was a baby then, and then, as now, their mother had spent her days in the mills. To Johnny had fallen the part of little father and little mother as well.

Will seemed to show the benefit of the giving over and the giving way. He was well-built, fairly rugged, as tall as his elder brother and even heavier. It was as though the life-blood of the one had been diverted into the other's veins. And in spirits it was the same. Johnny was jaded, worn out, without resilience, while his younger brother seemed bursting and spilling over with exuberance.

The mocking chant rose louder and louder. Will leaned closer as he danced, thrusting out his tongue. Johnny's left arm shot out and caught the other around the neck. At the same time he rapped his bony fist to the other's nose. It was a pathetically bony fist, but that it was sharp to hurt was evidenced by the squeal of pain it produced. The other children were uttering frightened cries, while Johnny's sister, Jennie, had dashed into the house.

He thrust Will from him, kicked him savagely on the shins, then reached for him and slammed him face downward in the dirt. Nor did he release him till the face had been rubbed into the dirt several times. Then the mother arrived, an anaemic whirlwind of solicitude and maternal wrath.

"Why can't he leave me alone?" was Johnny's reply to her upbraiding. "Can't he see I'm tired?"

"I'm as big as you," Will raged in her arms, his face a mass of tears, dirt, and blood. "I'm as big as you now, an' I'm goin' to git bigger. Then I'll lick you—see if I don't."

"You ought to be to work, seein' how big you are," Johnny snarled. "That's what's the matter with you. You ought to be to work. An' it's up to your ma to put you to work."

"But he's too young," she protested. "He's only a little boy."

"I was younger'n him when I started to work."

Johnny's mouth was open, further to express the sense of unfairness that he felt, but the mouth closed with a snap. He turned gloomily on his heel and stalked into the house and to bed. The door of his room was open to let in warmth from the kitchen. As he undressed in the semi-darkness he could hear his mother talking with a neighbour woman who had dropped in. His mother was crying, and her speech was punctuated with spiritless sniffles.

"I can't make out what's gittin' into Johnny," he could hear her say. "He didn't used to be this way. He was a patient little angel.

"An' he is a good boy," she hastened to defend. "He's worked faithful, an' he did go to work too young. But it wasn't my fault. I do the best I can, I'm sure."

Prolonged sniffling from the kitchen, and Johnny murmured to himself as his eyelids closed down, "You betcher life I've worked faithful."

The next morning he was torn bodily by his mother from the grip of sleep. Then came the meagre breakfast, the tramp through the dark, and the pale glimpse of day across the housetops as he turned his back on it and went in through the factory gate. It was another day, of all the days, and all the days were alike.

And yet there had been variety in his life—at the times he changed from one job to another, or was taken sick. When he was six, he was little mother and father to Will and the other children still younger. At seven he went into the mills—winding bobbins. When he was eight, he got work in another mill. His new job was marvellously easy. All he had to do was to sit down with a little stick in his hand and guide a stream of cloth that flowed past him. This stream of cloth came out of the maw of a machine, passed over a hot roller, and went on its way elsewhere. But he sat always in one place, beyond the reach of daylight, a gas-jet flaring over him, himself part of the mechanism.

He was very happy at that job, in spite of the moist heat, for he was still young and in possession of dreams and illusions. And wonderful dreams he dreamed as he watched the steaming cloth streaming endlessly by. But there was no exercise about the work, no call upon his mind, and he dreamed less and less, while his mind grew torpid and drowsy. Nevertheless, he earned two dollars a week, and two dollars represented the difference between acute starvation and chronic underfeeding.

But when he was nine, he lost his job. Measles was the cause of it. After he recovered, he got work in a glass factory. The pay was better, and the work demanded skill. It was piecework, and the more skilful he was, the bigger wages he earned. Here was incentive. And under this incentive he developed into a remarkable worker.

It was simple work, the tying of glass stoppers into small bottles. At his waist he carried a bundle of twine. He held the bottles between his knees so that he might work with both hands. Thus, in a sitting position and bending over his own knees, his narrow shoulders grew humped and his chest was contracted for ten hours each day. This was not good for the lungs, but he tied three hundred dozen bottles a day.

The superintendent was very proud of him, and brought visitors to look at him. In ten hours three hundred dozen bottles passed through his hands. This meant that he had attained machine-like perfection. All waste movements were eliminated. Every motion of his thin arms, every movement of a muscle in the thin fingers, was swift and accurate. He worked at high tension, and the result was that he grew nervous. At night his muscles twitched in his sleep, and in the daytime he could not relax and rest. He remained keyed up and his muscles continued to twitch. Also he grew sallow and his lint-cough grew worse. Then pneumonia laid hold of the feeble lungs within the contracted chest, and he lost his job in the glass-works.

Now he had returned to the jute mills where he had first begun with winding bobbins. But promotion was waiting for him. He was a good worker. He would next go on the starcher, and later he would go into the loom room. There was nothing after that except increased efficiency.

The machinery ran faster than when he had first gone to work, and his mind ran slower. He no longer dreamed at all, though his earlier years had been full of dreaming. Once he had been in love. It was when he first began guiding the cloth over the hot roller, and it was with the daughter of the superintendent. She was much older than he, a young woman, and he had seen her at a distance only a paltry half-dozen times. But that made no difference. On the surface of the cloth stream that poured past him, he pictured radiant futures wherein he performed prodigies of toil, invented miraculous machines, won to the mastership of the mills, and in the end took her in his arms and kissed her soberly on the brow.

But that was all in the long ago, before he had grown too old and tired to love. Also, she had married and gone away, and his mind had gone to sleep. Yet it had been a wonderful experience, and he used often to look back upon it as other men and women look back upon the time they believed in fairies. He had never believed in fairies nor Santa Claus; but he had believed implicitly in the smiling future his imagination had wrought into the steaming cloth stream.

He had become a man very early in life. At seven, when he drew his first wages, began his adolescence. A certain feeling of independence crept up in him, and the relationship between him and his mother changed. Somehow, as an earner and breadwinner, doing his own work in the world, he was more like an equal with her. Manhood, full-blown manhood, had come when he was eleven, at which time he had gone to work on the night shift for six months. No child works on the night shift and remains a child.

There had been several great events in his life. One of these had been when his mother bought some California prunes. Two others had been the two times when she cooked custard. Those had been events. He remembered them kindly. And at that time his mother had told him of a blissful dish she would sometime make—"floating island," she had called it, "better than custard." For years he had looked forward to the day when he would sit down to the table with floating island before him, until at last he had relegated the idea of it to the limbo of unattainable ideals.

Once he found a silver quarter lying on the sidewalk. That, also, was a great event in his life, withal a tragic one. He knew his duty on the instant the silver flashed on his eyes, before even he had picked it up. At home, as usual, there was not enough to eat, and home he should have taken it as he did his wages every Saturday night. Right conduct in this case was obvious; but he never had any spending of his money, and he was suffering from candy hunger. He was ravenous for the sweets that only on red-letter days he had ever tasted in his life.

He did not attempt to deceive himself. He knew it was sin, and deliberately he sinned when he went on a fifteen-cent candy debauch. Ten cents he saved for a future orgy; but not being accustomed to the carrying of money, he lost the ten cents. This occurred at the time when he was suffering all the torments of conscience, and it was to him an act of divine retribution. He had a frightened sense of the closeness of an awful and wrathful God. God had seen, and God had been swift to punish, denying him even the full wages of sin.

In memory he always looked back upon that as the one great criminal deed of his life, and at the recollection his conscience always awoke and gave him another twinge. It was the one skeleton in his closet. Also, being so made, and circumstanced, he looked back upon the deed with regret. He was dissatisfied with the manner in which he had spent the quarter. He could have invested it better, and, out of his later knowledge of the quickness of God, he would have beaten God out by spending the whole quarter at one fell swoop. In retrospect he spent the quarter a thousand times, and each time to better advantage.

There was one other memory of the past, dim and faded, but stamped into his soul everlasting by the savage feet of his father. It was more like a nightmare than a remembered vision of a concrete thing—more like the race-memory of man that makes him fall in his sleep and that goes back to his arboreal ancestry.

This particular memory never came to Johnny in broad daylight when he was wide awake. It came at night, in bed, at the moment that his consciousness was sinking down and losing itself in sleep. It always aroused him to frightened wakefulness, and for the moment, in the first sickening start, it seemed to him that he lay crosswise on the foot of the bed. In the bed were the vague forms of his father and mother. He never saw what his father looked like. He had but one impression of his father, and that was that he had savage and pitiless feet.

His earlier memories lingered with him, but he had no late memories. All days were alike. Yesterday or last year were the same as a thousand years—or a minute. Nothing ever happened. There were no events to mark the march of time. Time did not march. It stood always still. It was only the whirling machines that moved, and they moved nowhere—in spite of the fact that they moved faster.

When he was fourteen, he went to work on the starcher. It was a colossal event. Something had at last happened that could be remembered beyond a night's sleep or a week's pay-day. It marked an era. It was a machine Olympiad, a thing to date from. "When I went to work on the starcher," or, "after," or "before I went to work on the starcher," were sentences often on his lips.

He celebrated his sixteenth birthday by going into the loom room and taking a loom. Here was an incentive again, for it was piece-work. And he excelled, because the clay of him had been moulded by the mills into the perfect machine. At the end of three months he was running two looms, and, later, three and four.

At the end of his second year at the looms he was turning out more yards than any other weaver, and more than twice as much as some of the less skilful ones. And at home things began to prosper as he approached the full stature of his earning power. Not, however, that his increased earnings were in excess of need. The children were growing up. They ate more. And they were going to school, and school-books cost money. And somehow, the faster he worked, the faster climbed the prices of things. Even the rent went up, though the house had fallen from bad to worse disrepair.

He had grown taller; but with his increased height he seemed leaner than ever. Also, he was more nervous. With the nervousness increased his peevishness and irritability. The children had learned by many bitter lessons to fight shy of him. His mother respected him for his earning power, but somehow her respect was tinctured with fear.

There was no joyousness in life for him. The procession of the days he never saw. The nights he slept away in twitching unconsciousness. The rest of the time he worked, and his consciousness was machine consciousness. Outside this his mind was a blank. He had no ideals, and but one illusion; namely, that he drank excellent coffee. He was a work-beast. He had no mental life whatever; yet deep down in the crypts of his mind, unknown to him, were being weighed and sifted every hour of his toil, every movement of his hands, every twitch of his muscles, and preparations were making for a future course of action that would amaze him and all his little world.

It was in the late spring that he came home from work one night aware of unusual tiredness. There was a keen expectancy in the air as he sat down to the table, but he did not notice. He went through the meal in moody silence, mechanically eating what was before him. The children um'd and ah'd and made smacking noises with their mouths. But he was deaf to them.

"D'ye know what you're eatin'?" his mother demanded at last, desperately.

He looked vacantly at the dish before him, and vacantly at her.

"Floatin' island," she announced triumphantly.

"Oh," he said.

"Floating island!" the children chorussed loudly.

"Oh," he said. And after two or three mouthfuls, he added, "I guess I ain't hungry to-night."

He dropped the spoon, shoved back his chair, and arose wearily from the table.

"An' I guess I'll go to bed."

His feet dragged more heavily than usual as he crossed the kitchen floor. Undressing was a Titan's task, a monstrous futility, and he wept weakly as he crawled into bed, one shoe still on. He was aware of a rising, swelling something inside his head that made his brain thick and fuzzy. His lean fingers felt as big as his wrist, while in the ends of them was a remoteness of sensation vague and fuzzy like his brain. The small of his back ached intolerably. All his bones ached. He ached everywhere. And in his head began the shrieking, pounding, crashing, roaring of a million looms. All space was filled with flying shuttles. They darted in and out, intricately, amongst the stars. He worked a thousand looms himself, and ever they speeded up, faster and faster, and his brain unwound, faster and faster, and became the thread that fed the thousand flying shuttles.

He did not go to work next morning. He was too busy weaving colossally on the thousand looms that ran inside his head. His mother went to work, but first she sent for the doctor. It was a severe attack of la grippe, he said. Jennie served as nurse and carried out his instructions.

It was a very severe attack, and it was a week before Johnny dressed and tottered feebly across the floor. Another week, the doctor said, and he would be fit to return to work. The foreman of the loom room visited him on Sunday afternoon, the first day of his convalescence. The best weaver in the room, the foreman told his mother. His job would be held for him. He could come back to work a week from Monday.

"Why don't you thank 'im, Johnny?" his mother asked anxiously.

"He's ben that sick he ain't himself yet," she explained apologetically to the visitor.

Johnny sat hunched up and gazing steadfastly at the floor. He sat in the same position long after the foreman had gone. It was warm outdoors, and he sat on the stoop in the afternoon. Sometimes his lips moved. He seemed lost in endless calculations.

Next morning, after the day grew warm, he took his seat on the stoop. He had pencil and paper this time with which to continue his calculations, and he calculated painfully and amazingly.

"What comes after millions?" he asked at noon, when Will came home from school. "An' how d'ye work 'em?"

That afternoon finished his task. Each day, but without paper and pencil, he returned to the stoop. He was greatly absorbed in the one tree that grew across the street. He studied it for hours at a time, and was unusually interested when the wind swayed its branches and fluttered its leaves. Throughout the week he seemed lost in a great communion with himself. On Sunday, sitting on the stoop, he laughed aloud, several times, to the perturbation of his mother, who had not heard him laugh for years.

Next morning, in the early darkness, she came to his bed to rouse him. He had had his fill of sleep all the week, and awoke easily. He made no struggle, nor did he attempt to hold on to the bedding when she stripped it from him. He lay quietly, and spoke quietly.

"It ain't no use, ma."

"You'll be late," she said, under the impression that he was still stupid with sleep.

"I'm awake, ma, an' I tell you it ain't no use. You might as well lemme alone. I ain't goin' to git up."

"But you'll lose your job!" she cried.

"I ain't goin' to git up," he repeated in a strange, passionless voice.

She did not go to work herself that morning. This was sickness beyond any sickness she had ever known. Fever and delirium she could understand; but this was insanity. She pulled the bedding up over him and sent Jennie for the doctor.

When that person arrived, Johnny was sleeping gently, and gently he awoke and allowed his pulse to be taken.

"Nothing the matter with him," the doctor reported. "Badly debilitated, that's all. Not much meat on his bones."

"He's always been that way," his mother volunteered.

"Now go 'way, ma, an' let me finish my snooze."

Johnny spoke sweetly and placidly, and sweetly and placidly he rolled over on his side and went to sleep.

At ten o'clock he awoke and dressed himself. He walked out into the kitchen, where he found his mother with a frightened expression on her face.

"I'm goin' away, ma," he announced, "an' I jes' want to say good-bye."

She threw her apron over her head and sat down suddenly and wept. He waited patiently.

"I might a-known it," she was sobbing.

"Where?" she finally asked, removing the apron from her head and gazing up at him with a stricken face in which there was little curiosity.

"I don't know—anywhere."

As he spoke, the tree across the street appeared with dazzling brightness on his inner vision. It seemed to lurk just under his eyelids, and he could see it whenever he wished.

"An' your job?" she quavered.

"I ain't never goin' to work again."

"My God, Johnny!" she wailed, "don't say that!"

What he had said was blasphemy to her. As a mother who hears her child deny God, was Johnny's mother shocked by his words.

"What's got into you, anyway?" she demanded, with a lame attempt at imperativeness.

"Figures," he answered. "Jes' figures. I've ben doin' a lot of figurin' this week, an' it's most surprisin'."

"I don't see what that's got to do with it," she sniffled.

Johnny smiled patiently, and his mother was aware of a distinct shock at the persistent absence of his peevishness and irritability.

"I'll show you," he said. "I'm plum' tired out. What makes me tired? Moves. I've ben movin' ever since I was born. I'm tired of movin', an' I ain't goin' to move any more. Remember when I worked in the glass-house? I used to do three hundred dozen a day. Now I reckon I made about ten different moves to each bottle. That's thirty-six thousan' moves a day. Ten days, three hundred an' sixty thousan' moves. One month, one million an' eighty thousan' moves. Chuck out the eighty thousan'"—he spoke with the complacent beneficence of a philanthropist—"chuck out the eighty thousan', that leaves a million moves a month—twelve million moves a year.

"At the looms I'm movin' twic'st as much. That makes twenty-five million moves a year, an' it seems to me I've ben a movin' that way 'most a million years.

"Now this week I ain't moved at all. I ain't made one move in hours an' hours. I tell you it was swell, jes' settin' there, hours an' hours, an' doin' nothin'. I ain't never ben happy before. I never had any time. I've ben movin' all the time. That ain't no way to be happy. An' I ain't going to do it any more. I'm jes' goin' to set, an' set, an' rest, an' rest, and then rest some more."

"But what's goin' to come of Will an' the children?" she asked despairingly.

"That's it, 'Will an' the children,'" he repeated.

But there was no bitterness in his voice. He had long known his mother's ambition for the younger boy, but the thought of it no longer rankled. Nothing mattered any more. Not even that.

"I know, ma, what you've ben plannin' for Will—keepin' him in school to make a book-keeper out of him. But it ain't no use, I've quit. He's got to go to work."

"An' after I have brung you up the way I have," she wept, starting to cover her head with the apron and changing her mind.

"You never brung me up," he answered with sad kindliness. "I brung myself up, ma, an' I brung up Will. He's bigger'n me, an' heavier, an' taller. When I was a kid, I reckon I didn't git enough to eat. When he come along an' was a kid, I was workin' an' earnin' grub for him too. But that's done with. Will can go to work, same as me, or he can go to hell, I don't care which. I'm tired. I'm goin' now. Ain't you goin' to say goodbye?"

She made no reply. The apron had gone over her head again, and she was crying. He paused a moment in the doorway.

"I'm sure I done the best I knew how," she was sobbing.

He passed out of the house and down the street. A wan delight came into his face at the sight of the lone tree. "Jes' ain't goin' to do nothin'," he said to himself, half aloud, in a crooning tone. He glanced wistfully up at the sky, but the bright sun dazzled and blinded him.

It was a long walk he took, and he did not walk fast. It took him past the jute-mill. The muffled roar of the loom room came to his ears, and he smiled. It was a gentle, placid smile. He hated no one, not even the pounding, shrieking machines. There was no bitterness in him, nothing but an inordinate hunger for rest.

The houses and factories thinned out and the open spaces increased as he approached the country. At last the city was behind him, and he was walking down a leafy lane beside the railroad track. He did not walk like a man. He did not look like a man. He was a travesty of the human. It was a twisted and stunted and nameless piece of life that shambled like a sickly ape, arms loose-hanging, stoop-shouldered, narrow-chested, grotesque and terrible.

He passed by a small railroad station and lay down in the grass under a tree. All afternoon he lay there. Sometimes he dozed, with muscles that twitched in his sleep. When awake, he lay without movement, watching the birds or looking up at the sky through the branches of the tree above him. Once or twice he laughed aloud, but without relevance to anything he had seen or felt.

After twilight had gone, in the first darkness of the night, a freight train rumbled into the station. When the engine was switching cars on to the side-track, Johnny crept along the side of the train. He pulled open the side-door of an empty box-car and awkwardly and laboriously climbed in. He closed the door. The engine whistled. Johnny was lying down, and in the darkness he smiled.

Name ____________ **Course** ____________ **Date** ____________

Chapter Questions

1. The story states, "There had never been a time when [the boy] had not been in intimate relationship with machines. Machinery had almost been bred into him, and at any rate he had been brought up on it." Analyze this statement. What does it say about the work conditions during the Industrial Revolution for children?

2. Describe the scene when the superintendent and the inspector walk through the factory. What is each man's main purpose during the scene? Why are the boys so nervous during this visit?

THE EXCURSION

Book Eighth: The Parsonage

William Wordsworth

In 1884, poet William Wordsworth wrote "The Excursion" about his travels in the English countryside. In this excerpt he explores the negative effects of a factory on the men, women, and children who work there. Wordsworth uses similes and metaphors to compare the factory to a pagan temple and the workers to priests. In pre-Christian Europe priests were sometimes sacrificed as part of religious ceremonies.

Disgorged are now the ministers of day;
And, as they issue from the illumined pile,
A fresh band meets them, at the crowded door—
And in the courts—and where the rumbling stream,
That turns the multitude of dizzy wheels,
Glares, like a troubled spirit, in its bed
Among the rocks below. Men, maidens, youths,
Mother and little children, boys and girls,
Enter, and each the wonted task resume
Within this temple, where is offered up
To Gain, the master idol of the realm,
Perpetual sacrifice.

Name ______________ **Course** ______________ **Date** ______________

Chapter Questions

1. Explain with examples three of the metaphors or similes Wordsworth uses in this poem.

2. Why do you think Wordsworth chose to compare the factory to a pagan temple and the workers to sacrificial priests?

JOHN HENRY

Author Unknown

The *American Work Ballad* emerged from European and African ballad traditions with the addition of a uniquely American predilection for work as heroic. Figures like Paul Bunyan or Pecos Bill are geographically diverse but have common characteristics. The hero liked the work and was very good at it, possessed supernatural or mythic abilities, and died at work or "disappeared into the sunset" when the work was no longer available. The ballads "John Henry" and "Joe Magarac," featured here, represent the Work Ballad ideal, a type of redeemer figure who sacrifices himself to the work and for the good of his fellow workers.

By the late 19th century thousands of African-Americans had migrated North seeking work in the region's burgeoning railroad networks. "John Henry" emerged from this community and is a true folk song in that it has no recognized writer. Though technologies involved in railroading created John Henry's job, it is a machine that now threatens to take it.

When John Henry was a little bitty baby
No bigger than the palm of your hand
His daddy looked down at Johnnie and said
Johnnie's going to be a steel driving man
Lawd, Lawdy
Johnnie's going to be a steel driving man

When John Henry was a little bitty baby
Sitting on his daddy's knee
Well, he picked up a hammer and a little piece of steel
He said This hammer's going to be the death of me
Lawd, Lawdy
This hammer's going to be the death of me!

Mr. Peabody bought a steam hammer.
He brought it to the end of the track.
Then he said bring out your best
We're gonna' have us a test
And if my hammer wins the rest of you can pack
Lawd Lawdy
If my hammer wins, the rest of you can pack.

John Henry Went to the tunnel to drive
Steam hammer by his side
He said Before I would let that steam hammer get me down
You know I'd lay down this hammer and I'd die
Lawd Lawdy
I'd lay down this hammer and I'd die

Now the straw boss came to John Henry
He cried This tunnel is caving in
John Henry just laughed at the strawboss and said,
That's nothing but my hammer sucking wind
Lawd, Lawdy,
That ain't nothing but my hammer sucking wind.

John Henry turned to the straw boss.
He said, "Lawd knows how hard I tried.
I done my best but that hammer is fast.
Please gimme a cool glass of water 'fore I die!
Lawd, Lawdy.
Gimme a cool glass of water 'fore I die!

Mr. Peabody wanted a measure
So they brought out the tape so fine
John Henry laid down 12 feet of cold hard track
The tell me that the hammer just laid 9
Lawd, Lawdy
They tell me that the hammer just laid 9.

John Henry had a sweet little wife
Her name was Polly Ann
When Johnnie got sick and
He had to go to bed
Poly drove that steel just like a man
Lawd, Lawdy
Poly drove that steel just like a man

John Henry died in the evening
Left his poor widow to cry
Well, they laid John Henry by the railroad track
So he could hear the trains go rumbling by
Lawd, Lawdy
So he could hear the trains go rumbling by.

John Henry was a steel driving man
He drove steel all over this land
And every time a train goes rumbling by his grave
They say 'Down yonder lies a Steel driving man!'
Lawd Lawdy
Yonder lies a steel driving man.

JOE MAGARAC

Randy Sparks

Although the origins of the Joe Magarac legend are steeped in mystery, the character represents East European immigrants who worked in Pittsburgh area steel mills in the late 19th and early 20th centuries. In the 1960s, Randy Sparks, leader of the folk music group The New Christy Minstrels, penned this ballad based on the story.

© Avatar_023/Shutterstock.com

To the old steel mill big Joe went down
Joe Magarac
He said all you hunkies gather round
Big Joe Magarac
I heard they're gonna build a railroad to Frisco and back
From Maine down to Mexico
And who's gonna make the steel for that track
Big Joe Magarac

Joe Magarac
Joe Magarac
Well there ain't no steel man in all this world like
Joe Magarac

From the Lackawanna Road to the Rock Island Line
went Joe Magarac
and the railroad boss said you're doing just fine
Big Joe Magarac

You know we gotta have some buildings that reach to the sky
And highways wherever you go
Well, I aint done yet and I'm sure gonna try
says Joe Magarac

Joe Magarac
Joe Magarac
Well there ain't no steel man in all this world like
Joe Magarac

Well the melter boss said we gotta shut down the mill for Joe Magarac
He done worked all the ore in the mine on the hill
Oh Joe Magarac
Joe jumped into the ladle it was fiery hot and he waved to his hunkies below
And the roller boss said the very best steel we got is Joe Magarac

Joe Magarac
Joe Magarac
He was made out of steel from his head to his toes
Big Joe Magarac

Name ______________________ Course ____________ Date _______________

Chapter Questions

1. What is the overall attitude toward the Industrial Revolution shown by these ballads and what textual evidence led to your answer?

2. What characteristics made John Henry and Joe Magarac typical American folk heroes similar to Paul Bunyan or Pecos Bill?

HONEST WORK

Todd Rundgren

By the late 20th century songwriters produced work ballads decidedly less optimistic about the worker and his or her place in society. They emphasized the soul-shattering alienation workers could experience in a globalized economy that seemed out of their control. Enigmatic singer-songwriter Todd Rundgren released "Honest Work" on his 1985 album *A Capella*. The song uses a traditional ballad form to explore the thoughts of a man left unemployed in the wake of changes in the workplace and in society at large. In the first half of the song the singer blames himself for not obtaining skills that could help him adapt to changing technology. By the second half of the song he has broadened the indictment to "the prophets of a brave new world," a world in which he has become "society's debris."

I'm not afraid to bend my back
I'm not afraid of dirt
But how I fear the things I do
For lack of honest work
My family is lost to me
They could not bear the hurt
To see the state their boy is in
For lack of honest work

I hold no blame for anyone
'twas I who did arrange
To pay my union dues so I'd

Not have to learn or change
And when I was replaced, 'twas I
Who started down the hill
And drank away my savings 'til
I couldn't stop myself

The prophets of a brave new world
Captains of industry

Have visions grand and great designs
But none have room for me
They see a world where everyone
Is rich and smart and young
But if I live to see such things
Too late for me they come

I know I'm not the only one
To fall beneath the wheel
Such company can not assuage
The loneliness I feel
So many are resigned to be
Society's debris
But I will be remembered for
The life life took from me

For I'm not afraid to bend my back
I'm not afraid of dirt
But how I fear the things I do
For lack of honest work

Name ______________________ Course ____________ Date ______________

Chapter Questions

1. Whom does the singer blame for his troubles in the song's second half and why do you think he expanded the blame beyond himself?

2. Discuss the statement, "I know I'm not the only one to fall beneath the wheel." Specifically, what is "the wheel"?

YOUNGSTOWN

Bruce Springsteen

Bruce Springsteen is widely regarded as one of the foremost songwriters of the Rock and Roll era. Among traditional Rock themes of rebellion, love, and escapism Springsteen has written entire albums dedicated to the plight of the "common man." This song appears on the 1995 album The Ghost of Tom Joad and is set in the steel mill town of Youngstown, Ohio. The singer is alienated from his work and the products of that work but also feels a social alienation when comparing his life experiences to those of his father. His father was proud of the mill and its product and proud of the "good" war (World War II) which America won with its steel and industrial might. By contrast, the singer works only for his paycheck and fought in the unpopular Vietnam War.

© Steve Broer/Shutterstock.com

Here in northeast Ohio
Back in eighteen-o-three
James and Dan Heaton
Found the ore that was linin' Yellow Creek
They built a blast furnace
Here along the shore
And they made the cannonballs
That helped the Union win the war

Here in Youngstown
Here in Youngstown
My sweet Jenny I'm sinkin' down
Here darlin' in Youngstown

Well my daddy worked the furnaces
Kept 'em hotter than hell
I come home from 'Nam worked my way to scarfer
A job that'd suit the devil as well
Taconite coke and limestone
Fed my children and make my pay
Them smokestacks reachin' like the arms of God
Into a beautiful sky of soot and clay

Here in Youngstown
Here in Youngstown
Sweet Jenny I'm sinkin' down
Here darlin' in Youngstown

Well my daddy come on the Ohio works
When he come home from World War Two
Now the yard's just scrap and rubble
He said "Them big boys did what Hitler couldn't do."
These mills they built the tanks and bombs
That won this country's wars
We sent our sons to Korea and Vietnam
Now we're wondering what they were dyin' for

Here in Youngstown
Here in Youngstown
My sweet Jenny I'm sinkin' down
Here darlin' in Youngstown

From the Monongahela valley
To the Mesabi iron range
To the coal mines of Appalachia
The story's always the same
Seven hundred tons of metal a day
Now sir you tell me the world's changed
Once I made you rich enough
Rich enough to forget my name

And Youngstown
And Youngstown
My sweet Jenny I'm sinkin' down
Here darlin' in Youngstown

When I die I don't want no part of heaven
I would not do heaven's work well
I pray the devil comes and takes me
To stand in the fiery furnaces of hell

Name ____________________ Course ____________ Date ______________

Chapter Questions

1. Discuss how the singer is alienated from his work.

2. Why do you think he and his father viewed the steel mill and its products so differently?

THE WORLD IS FLAT

Thomas L. Friedman

In the introduction to his national bestselling book, The World is Flat, three time Pulitzer Prize winning journalist, Thomas Freidman, uses the title as a metaphor for the world which he believes is now a level playing field regarding commerce. Distance and time are no longer barriers to free trade. How did the fall of the Berlin Wall open up new trade opportunities? Did Netscape make computing revolutionary? In the book's introduction, printed here, Friedman discusses ten 'flatteners' that have changed the economic horizon and provided unlimited access to previously untapped markets and labor.

© haveseen/Shutterstock.com

Let me give you the brief thesis. The first chapter of the book is called, appropriately, "While You Were Sleeping." As I pointed out in the *Times Magazine* the other day, the main argument here is that Christopher Columbus set sail in 1492 looking for, as we all know, a shorter route to India. That is where he was going. He had the *Nina*, the *Pinta*, and the *Santa Maria*. He never did find India, but he called the people he met Indians, nevertheless. He came home and told his wife, "Honey, I've accidentally discovered that the world is round."

I set off for India 512 years later. I knew just which direction I was going. I went east. I had Lufthansa business class and a GPS satellite that popped up in my seat. I came home and reported to my wife—and only my wife, and only in a whisper—"Honey, I've discovered accidentally that the world is flat."

The first chapter is really about all the encounters I had, first in India and then beyond. I kept going east, to Dalian, China, which is now the outsourcing capital of Japan. Japanese-speaking Chinese, by the tens of thousands, arc now doing the backroom operations for many global companies, including G.E. and Microsoft, from Dalian, China, for the Japanese market.

"The World is Flat" by Thomas L. Friedman. Carnegie Council for Ethics in International Affairs, www.carnegiecouncil.org By permission.

Then I kept going east, to the McDonald's franchises that are in a pilot operation, where if you drive up to the McDonald's now and you say, "I want three milkshakes and four Big Macs and eighteen fries," you are actually talking to someone in Colorado—you are not talking to someone at that McDonald's—who is taking down your order electronically, and taking your picture, incidentally, and zapping your picture and your order to a processor inside that McDonald's, who actually prepares the food and hands it to you, corresponding to your picture. McDonald's discovered they saved thirty seconds on every order when they did that and drove their error rate down to massively low levels.

I kept going east. I called JetBlue to make a reservation. I knew what was going on. After I tried to make my reservation, I said to the woman on the phone, "Ma'am, could I ask you your name?" She said, "Sure. My name is Betty." I said, "Betty, where are you right now?" She said, "Honey, I'm up in my bedroom in Salt Lake City. I'm in my slippers and my bathrobe, and I'm looking out at this beautiful winter scene." David Neeleman, the founder of JetBlue, has set up a reservation system that basically is people—housewives and retirees—in Salt Lake City. If you call JetBlue, that is who you are talking to. It is a completely home-based reservation system.

As I kept moving east and seeing these flattening forces, I, finally ended up back home in Washington, D.C.

Now, the basic meta-argument I make in this first chapter is that there have been three great eras of globalization. One I call Globalization 1.0. It was from 1492 to about 1800. It shrunk the world from a size large to a size medium. That era of globalization was built primarily around countries globalizing. The agent of globalization was Spain discovering America, Great Britain colonizing India, and Portugal in the East Indies and East Asia.

The second great era of globalization, I would argue, is around 1820, 1825—that was really the beginning of global arbitrage—until the year 2000. That era shrunk the world from a size medium to a size small. That era of globalization was built around companies globalizing. That was the dynamic agent, companies globalizing for markets and for labor.

What I am arguing in this book is that while you were sleeping—certainly while I was sleeping—we actually entered Globalization 3.0, which began in the year 2000—by accident, right at the turn of the millennium. This era is shrinking the world from a size small to size tiny, and leveling and flattening the playing field at the same time.

What is unique about this era is that it is not built around countries, and it is not built around companies. It is built around individuals, The new "new" thing about this era of globalization is that the dynamic agent is the individual and the small group's ability to globalize. Be advised, it is not going to be a bunch of white Western individuals. It is going to be individuals of all colors of the rainbow.

The second chapter of the book is the longest chapter. It is a third of the book. It is the one that took me the most time and really forced me to retool myself. It is a whole chapter devoted to answering that question that burned in me when I was doing those interviews in India. That question was, how did this happen?

That chapter is called "The Ten Days That Flattened the World"—it is really "The Ten Forces That Flattened the World." This is my attempt to explain what the forces were that created this level playing field. I have come up with ten. Let me go through them very quickly.

Day 1: The first flattener is 11/9—not 9/11, 11/9. In a wonderful cabalistic accident of dates, the Berlin Wall came down on 11/9, November 9, 1989. The fall of the Wall was a huge flattener, When the Wall was there, it was very hard—in fact, it was impossible—to actually see the world as a single flat space. But when the Wall was removed, suddenly you could see from one end of the world, metaphorically, to the other.

I daresay if you did a LexisNexis search of the term "globalization," before the Berlin Wall, you would find it rarely used, because, after all, who could have a global policy? You could have an Eastern policy, you could have a Western policy, but it was very hard to have a global policy when there was a wall in the way. So the fall of the Wall was hugely important for us to perceptually see and think of the world as a single space.

What I call this first flattener is, "When the Wall came down and the Windows came up," Windows operating system 3.0 shipped six months after the fall of the Berlin Wall. Those two things happened at the same time, giving us a universal graphical user interface from one end of the world to the other, to look at the world through. That was the first flattener.

Day 2: The second flattener occurred on 8/9/95, which I consider to be as important a day in our lifetime as the fall of the Berlin Wall. 8/9/95, August 9, 1995, is the day a small company in Mountain View, California, called Netscape, went public. Netscape's IPO was a transformative day in our lifetime, for three reasons:

1. Netscape going public gave us the Internet browser. The browser is that thing that allows us to illuminate and illustrate all the data, information, words, and music that are in those Web sites. It is what allowed us to browse the Internet and visually see what was there. It was a huge breakthrough, an invention by Mark Andreessen from the University of Illinois. That going public basically made the Internet open and accessible to everyone from grandma and grandpa to a five-year-old grandchild. It essentially gave us the Internet.
2. It gave us—this is a little more technical—a set of transmission protocols, kind of digital pipes, that made the Internet interoperable. You may recall, when you first started using the Internet, that you were on CompuServe and your wife was on AOL, and you had a hard time communicating. What Netscape did was, basically, commercialize a set of open standards that really made the Internet interoperable.
3. Third and most importantly, Netscape triggered the dot-com boom, which triggered the dot-com bubble, which triggered the crazy, wild, ridiculous, massive overinvestment in

fiberoptic cable—$1 trillion in five years. Netscape came out at $28.00. It opened at $71.00. It closed the first day at $56.00. People all over America looked at that and said, "There is gold over them there hills." That started the dot-com boom. That prompted every yahoo from Global Crossing to Lucent, to your local telephone company, to think that the demand for bandwidth is going to be infinite, and if the demand is going to be infinite, you can't lay down enough fiberoptic cable under the ocean and around the world. So every one of these companies tried to do just that. You and I, with our investment dollars, paid for it. The accidental result was that within five years of Netscape going public, Bangalore and Beijing and the Bronx had become next-door neighbors. We drove down the cost of transmitting words and music and data between all these points to zero. What the Netscape moment did, in sum, was bring people-to-people connectivity. People were able to connect with people elsewhere like never before.

Day 3: The next flattener I simply call workflow. It is really my summary of all the software and standards—standards are very important—that connected all that bandwidth from those fiber-optic cables with all those PCs. Workflow is very important. Again, if you go back to the early 1990s, you may recall that the Carnegie Council here—their inventory department was running Microsoft Windows, but their membership department was running Novell. Each department was more efficient, because they were working on computers. There was just one problem: Their Novell computers couldn't talk to their Windows computers. So they were never as productive as they could be.

There was a revolution going on in the late 1990s, thanks to all kinds of digital pipes and things like middleware, that enabled applications to talk to applications. Suddenly, everyone's inventory department could talk to everyone's sales department without ever thinking about it. More importantly, Carnegie's sales department and inventory department could talk to the UN's inventory and sales departments, no matter what machine they were operating on and no matter what software they were running. That was a huge, huge revolution, because when applications could talk to applications, suddenly we could work together in whole new ways.

After these first three flatteners, we get what I call the "genesis" of the flat world. It happened right at the end of the 1990s when, suddenly, people could talk to people like never before and, at the same time, applications could talk to applications like never before. Put those two together and, willy-nilly, you have a global platform for multiple forms of collaboration. Suddenly, people can collaborate on more kinds of work, in more different ways, on more different days, and from more different places than ever before. The next six flatteners are the new forms of collaboration that immediately sprung from this platform.

Day 4: The first new form of collaboration was **outsourcing.** Suddenly, I can take the Carnegie Council's accounting department and move it to North Dakota, to Arizona, or to Bangalore, because when people-to-people and application-to-application can connect, I can move it anywhere—where the most efficient, effective, or cheapest producer can operate it.

© Mathias Rosenthal/Shutterstock.com

Day 5: The second new form of collaboration was what I simply call **offshoring**. This existed before, but it has gone to a whole new level, basically, thanks to the flattening of the world. Off-shoring is when I take my factory from Canton, Ohio, and move it lock, stock, and barrel to Canton, China, where it then gets integrated into my whole production operation. Offshoring dates from when China joined the World Trade Organization and from when India began to remediate all our computers for Y2K.

Day 6: The third great new form of collaboration is **open sourcing**. Now, **I** have a bunch of geeks sitting at home working for free, writing the next operating system, called Linux. How would you like to be Bill Gates? Your whole business model all these years—you are a tough businessman—is that anybody who comes up and tries to compete with you, you undercut them, because you are big and tough and you are Microsoft. Then one day you discover that your main competitor—Linux now has 15 percent of the operating-system market—is selling your product for free. It is kind of hard to undercut that, especially when all your other competitors are funding people to work on it for free, so they can get into your cash hoard. This is a whole new model of production. Some of you, I am sure, have kids in college. Ask them how many of them have downloaded Firefox. Firefox is a new browser. As we all know, the browser on all our computers is Internet Explorer, made by Microsoft. Through a very interesting collaboration last year, a nineteen-year-old at Stanford, working with a twenty-four-year-old in New Zealand, never having met face to face, collaborated in an open-source forum. The new browser they created was downloaded 10 million times in the first month. Firefox now has 5 percent of the browser market. In the book, I profile a remarkable young man, Brian Behlendorf, who was the leader of the Apache movement. All your computers basically operate on a Web server. The Web server is what goes to that Web site when you tell it to, to get the information there. Every Web site has to have a Web server to serve it up to your computer, The Web server that really runs 75 percent of the computers in America is called Apache. The chief of the Apache tribe is a young man named Brian Behlendorf. These people, in an open-source forum, produced a Web server that was so good, it basically prompted IBM to throw out its Web server and pay its own engineers to join the open-source chat room of Apache that produced today's standard Web server.

Open sourcing, a whole new form of collaboration. It is not just about Web servers. I use Wikipedia constantly. I don't know how many of you have used Wikipedia. It is an open-source dictionary and encyclopedia, where people simply contribute the entries on their own. It is self-corrective. If you have never dipped into the open-source world, I urge you to do so. It is going to be, I think, the main post-industrial forum of creative innovation.

Day 7: The fourth new form of collaboration I call **supply-chaining**. "Supply-chaining" is simply my word for Wal-Mart. Wal-Mart has designed a global supply chain down to the last atom of efficiency, so if you take an item off the shelf in Brooklyn, another is immediately made in Shenzhen, China. As you may have heard, if Wal-Mart were a country, it would be China's eighth-largest trading partner today.

Day 8: The fifth new form of collaboration that sprung off this platform I call **insourcing**. I had the most fun learning about this, actually. Insourcing is what UPS does. If you think the folks in the funny brown shorts in the funny brown trucks are just delivering packages, then you have not been paying attention to the little new logo on the side of that funny brown truck. You know what it says now? It doesn't say, "Packages delivered." It says, "Your world synchronized." What UPS does now is go inside your company—it is why I call it insourcing—and take over your whole internal logistics operation right up to your neck.

My favorite example: If you have a Toshiba laptop and your Toshiba laptop breaks, what do you do? You look at the warranty. It says, "Call 1-800-HELP." So you call 1-800-HELP, and when you call there, they tell you to take your Toshiba laptop down to the UPS store and send it to Toshiba, and they will repair it and send it back. What you don't know is that your Toshiba laptop goes from the UPS store to the UPS hub at Louisville Airport in Kentucky, their global headquarters, where in a clean room, in a hangar, at Louisville Airport, your Toshiba laptop is repaired by a UPS employee. Your Toshiba laptop never goes back to Toshiba. They don't want to see it. They insource all of it to UPS.

You call nike.com, go online to order a pair of sneakers for your kids. UPS answers that email. UPS picks and packs the shoes. UPS ships them. UPS bills them, and UPS collects the money. You see the Papa John's Pizza truck go by. Guess who is driving? It ain't Papa John's. It is someone in funny brown shorts working for UPS.

There are whole companies today—and you would be shocked to know how many—who never touch their products anymore. They have been completely insourced to UPS or FedEx or DHL. This is a huge flattener going on. To do this requires an enormous number of standards, in order to make this operate at a seamless level.

Day 9: The last new form of collaboration I call **informing**. Informing is what Google, Yahoo, and Microsoft Search do. They allow you to collaborate now with data all by yourself. I had a research assistant, my assistant in Washington, who worked with me on this book. But my real research assistant for this book was a person named Google. I was able to mine all my data all by myself. I was able to inform myself. Inform me: that is what Google does; it is what TiVo does. It allows you to inform and entertain yourself—a whole new form of collaboration.

Day 10: The tenth I simply call the "**steroids**." The steroids are voice-over-the-Internet and wireless. What the steroids are doing is turbo-charging all six of these new forms of collaboration and now allowing you to do any one from anywhere with any device.

So those are the ten days that flattened the world. You will understand why it took me a long time (a) to figure that out, and (b) to go down to UPS and Wal-Mart and do all the research around this, so that I was able to understand them enough to explain it to myself and, hopefully, to you.

The third chapter of the book is really the core thesis. It is called "The Triple Convergence." The triple convergence is basically what, I argue, is the pivot point for the flattening of the world. The first convergence was that sometime around 2000, all ten of these flatteners started to converge. The complementarities between them all suddenly started to work together. The informing drove the outsourcing; the outsourcing drove the insourcing; the insourcing drove the supply-chaining; the supply-chaining drove the offshoring. All started to work together at a tipping point, and the result was a Web-enabled global platform for multiple forms of collaboration and sharing of knowledge—irrespective of distance, of geography, and increasingly, even of language. We are just at the beginning of this.

The second convergence—as I said, it was a triple convergence—was the fact that we are now at the beginning of learning to what I call "horizontalize" ourselves. We are going from a world where value was created almost exclusively in command-and-control silos, from the top down, to a world where value is going to increasingly be created by connect-and-collaborate, horizontally. We are going from a vertical value-creation model to a horizontal value-creation model. To get the most out of this, you have to change all your habits.

The Stanford economist Paul David wrote a famous essay on electrification. He asked, when electricity came out, why didn't we get a burst of productivity? His answer was, because people had to change all their habits. They had to change the very architecture of buildings in order to accommodate small electric motors. It was only after twenty years, when everyone changed their habits from steam engines and their architecture and their business-management processes, did we get the explosion of productivity from electrification.

I would argue the same is going to have to happen with "horizontalization," as we move from vertical to horizontal. An example I like to give from this—and it is a very vivid one—happened to me when I was starting to write this book. My daughter goes to school in New Haven. I live in Bethesda, Maryland. Last spring, I was going to visit her, taking her some spring clothes. To get from Bethesda, Maryland, to New Haven is a total pain in the behind. You have to drive to BWI

Airport in Baltimore, take Southwest Airlines to Hartford, and then drive an hour from Hartford to New Haven.

You don't look like a Southwest crowd to me, but never mind. If you have flown Southwest, you may know that on Southwest Airlines you don't get an assigned seat; you just get a ticket that says A, B, or C. You do not want to be a C on Southwest Airlines. You don't even want to be a B if you are carrying two bags of carry-on for your daughter and want to find room in the bin overhead or not get stuck in the middle seat.

No problem. I am a hip guy. I did the e-ticket thing. I got to Southwest Airlines ninety-five minutes before my flight. I put my credit card in the Southwest e-ticket machine. Out came my ticket, and it said B. I said, "This thing is rigged! This is fixed! This is worse than Las Vegas! There is no way I'm a B! I'm here ninety-five minutes before this flight. There is no way I'm a B!" I kicked the thing, went and got my Cinnabon and stewed in the back of the B line. So I was sitting at the gate for forty-five minutes in the back of the B line. Then they called the flight. Then I saw it. All the A's were getting on with what looked to me like crumpled pieces of white printer paper, as if they had downloaded their boarding passes at home and printed them out on their home computers.

What, of course, I didn't know was that Southwest Airlines had just instituted a program where at 12:01 A.M. the night before, because of the convergence, that first convergence, they could make their customers their employees. They could get you to stay up until 12:01 A.M. the night before, and if you valued your time, in effect, pay Southwest Airlines to be their employee and download your boarding pass and your barcode on your own home printer paper the night before. I looked at that, friends, and I said, "Friedman, you are so twentieth-century. You are so Globalization 2.0."

Think about it. In our lifetime, we went from a ticket agent, in Globalization 1.0, a physical person who stood before you, to the e-ticket machine, to you becoming the ticket agent. You just saw us go from Globalization 1.0, to 2.0, to 3.0, in a decade. That transition is happening in every business. I can't tell you how many CEOs I spoke to who all told me the same thing, like they were preprogrammed: "I've had to do the most amazing thing in order to stay alive in this business."

That is the second convergence. I had to learn to "horizontalize" myself. I had to learn to collaborate totally differently with Southwest Airlines to get the productivity boost for myself and for Southwest Airlines. We are just at the beginning of that.

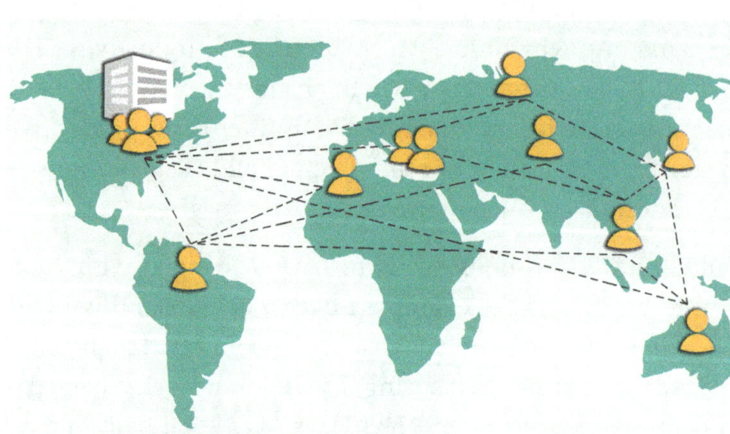

© Makkuro GL/Shutterstock.com

The third convergence—I said there were three—is that

right when the world was flattened, what accidentally happened was that 3 billion people who were out of the game walked onto the playing field—from India, China, and the former Soviet empire. When do they walk onto the playing field, friends? Just when it has been flattened. Just when they can plug and play, connect and collaborate, more directly with your kids and mine than ever before in the history of the planet.

Oh, there are 3 billion of them, but only 5 percent can plug and play. Excuse me, that is 150 million by my math. That is the exact size of the American workforce. Also, they can plug and play much more quickly, because they have no embedded legacy systems. That is why there are more cell phones in China today than there are people in America. They can go straight to the latest technology.

The simple thesis of this book is that the triple convergence of this flattened world with these new businesses processes, with these 3 billion new players, is defining the course of the twenty-first century. Let me simply close by saying that this triple convergence happened to coincide with a political "perfect storm." The political perfect storm was called Enron. Enron, 9/11, and the dot-com bust made a lot of morons think and write that globalization was over. Actually, just the opposite was happening. 9/11 completely distracted us—most of all, myself—from everything that was going on. Enron made it so that every business executive was guilty until proven innocent; therefore, none of them wanted to talk about any of this.

When I went around and interviewed people for this book, it was a bit like being in a science fiction movie. They are like pod people, all these CEOs. I learned everything that I know for this book from them. They know just what is going on. But nobody has told the kids. Nobody is talking about this. So we are at one of the great inflection points, I would argue, in our history. I think this is going to be as big as Gutenberg when it plays out. And nobody is talking about it.

Name ______________________ Course ____________ Date _______________

Chapter Questions

1. Briefly describe three of the innovations that the author says "flattened" the world.

2. How do these innovations function to enable forms of collaboration today, or do they? Explain.

3. Are there other ways that globalization is being spread in today's economy?

4. Among these ten things that the author says "flattened" the world, choose two and discuss what happened in your own words.

INTERNATIONAL FORUM ON GLOBALIZATION: WHAT SHOULD BE OFF-LIMITS TO GLOBALIZATION?

The International Forum on Globalization is a group comprised of economists, scholars, activists, and others concerned about the long-range impact of economic globalization.

© Anton Balazh/Shutterstock.com

As recently as two decades ago, large parts of the world were not part of economic globalization. The majority of people in the world still lived off the land, many with little dependence on outside markets. In many rural areas, seeds were exchanged as the collective property of the community, not the private property of Monsanto or Cargill. Many of the three hundred million indigenous people in the world lived in complete isolation from global trade activity. Most municipal water systems were under local government or community control. Much of the economic activity in the Soviet Union, Eastern Europe, and China was not linked to global markets. Most developing countries restricted foreign investment in their banking, insurance, and other critical economic sectors. Most stock markets were national, closed to global investors. Even though global corporations clamored to enter each of these domains, national and local governments and communities maintained strong barriers.

All of that has changed. Under two decades of market fundamentalism, introduced by Ronald Reagan, Margaret Thatcher, Helmut Kohl, and their counterparts elsewhere, the boundaries came crashing down. Some of this was seen in dramatic fashion on CNN in living rooms around the world, such as the destruction of the Berlin Wall. Some happened in the face of remarkable citizen opposition, such as the passage of NAFTA in 1993 and the WTO in 1994. Other battles over the global spread of corporate control occurred on the local stage, such as the determined fight by Bolivian workers and peasants to keep the municipal water system in Cochabamba out of the hands of Bechtel and the struggles of Indian peasants against the Cargill and Monsanto assertions of property rights over their seeds. During these two decades, global corporations—with the strong support of many national governments—forcefully asserted their right to any market anywhere. And today their reach has extended into virtually every domain of even remote rural communities around the world.

One of the most critical points of unity among the authors of this document is that this encroachment of corporate globalization into every aspect of life and the environment must stop. We seek to shift the framework of the overall debate on globalization in this sense: we believe that many aspects of social and economic life around the world should be off-limits to the processes of economic globalization. In this chapter, we offer the beginnings of a framework for choosing which arenas should be off-limits to which aspects of economic globalization.

On one level, there is already agreement across the spectrum of the globalization debate that certain goods and services should be kept out of trade. For example, governments around the world have created a global convention to ban trade in hazardous wastes. Likewise, there is a global convention against the trafficking of endangered species. And there is growing global action against the trafficking across borders of women sold into sexual bondage.

Now, the International Forum on Globalization would like to expand the debate beyond the "pernicious" goods mentioned above to include the rights of peoples and obligations of nations concerning what has traditionally been called *the commons*.

In this chapter, we offer an overview of the many notions of the commons around the world and spell out the current threats of economic globalization to commons such as freshwater, the genetic commons, communal lands, and others. We then introduce the concept of the modern commons—that is, the role of governments in carrying out a sacred public trust to perform certain key services that were once the province of communities and families but have been captured by and subsumed into the nation-state. We argue that selling off these services to global corporations—which operate on an entirely different set of priorities than the public interest—is a grave violation of these modern commons, many of which should never be commodified. We argue that they are also obligations of governments as trustees of the common rights and services of people. Finally, we offer a few ideas on how the commons might be protected from the worst aspects of economic globalization.

We offer these suggestions in the spirit of opening up a complex discussion, We do not pretend to present the final answers.

UNDERSTANDING THE COMMONS

Much of the thrust of economic globalization over these decades has been driven by global corporations pushing to develop and market every type of natural resource. In a world where natural resources were already seriously overexploited, corporations have attempted to convert every remaining nook and cranny of the natural world and human experience into commodified form.

Now, areas of life traditionally considered out of bounds are being considered for monetized activity, private ownership, or global trade. These are aspects of life that had been accepted since time immemorial as collective property, or the common heritage of all peoples and communities, existing for everyone to share as they have for millennia. These are what have been known as the commons.

Obvious among them are the air we breath, the freshwater we drink, the oceans and the diverse wildlife and plant biodiversity of the world, the genes all creatures pass to following generations, the stores of human knowledge and wisdom, the informal support systems of the community, the seeds that communities use for replanting, the public square, shared languages and culture, and among indigenous peoples, communal lands that have been worked cooperatively for thousands of years.

Some commons are gifts from the bounty of nature and are crucial to the survival of people and the earth. Most cultures have innumerable rituals to celebrate these gifts and rules or taboos against harming them. Other commons are new, including, for example, the broadcast spectrum or the Internet. Still others are ancient, such as the common grazing meadows of Africa, Europe, and Asia, folklore, and cultural artifacts.

Some commons may be thought of as global, such as the atmosphere, the oceans, outer space, and because they have no territorial claimants, Antarctica and the moon. Others may be thought of as community commons: public spaces, common lands, forests, the gene pool, local innovative knowledge with respect to medicinal plants, and seeds that communities have developed over centuries.

Author Jonathan Rowe of California's Tomales Bay Institute points out that the key characteristic of all aspects of the commons is that they belong to everyone. No one has traditionally had exclusive rights to them. We have inherited them jointly; they are our common heritage. They "are more basic to our lives than the state or the market," says Rowe. He goes on to say, "One cannot imagine a life without air fit to breathe, oceans rich with life, free clean water, a vibrant biodiversity. These are things we have always taken for granted. The commons have the quality of always having been there, one generation after another, available forever to all."

Later in this chapter, we will advocate for yet another category—the "modern commons" of public services like health, water purification and distribution, education, information, each of which was once achieved informally within small local and indigenous communities that have since been absorbed by the state and are also now on tap for privatization.

The modern nation-state has also taken onto itself the collective security of its citizens, which in a less technologically oriented and mobile world was once the province of communities. In that context, we need to discuss the varieties of security protection that a modern state is obliged to offer in addition to its obvious military roles. Since adopting the Universal Declaration of Human Rights in 1948, the United Nations has helped governments define basic human rights. Governments are obligated to protect human rights, as well as food security, as fundamental to life. Protection of cultural diversity is also a basic duty and right. Nothing in the global trading system should ever be permitted to reduce these fundamental priorities. (At present, many elements of the WTO and other trade agreements work directly against countries that try to protect these fundamental rights, which we now propose as part of the modern commons.)

Precise categorization of each type of commons is difficult to achieve because many cross several categories—for example, river water (which may pass through several regions and countries), biodiversity (which may be local or national), the broadcast spectrum (which may be local, national, or international), and the genetic structures of life. Similarly, protections against trade that destroys the commons—toxins, armaments, and so on—must be national and international.

Nonetheless, the purpose of this discussion is to lay on the table one central principle:

> Any global trading system needs to recognize and yield to the primary notion that not every aspect of experience should be subject to its centralized rules, and many aspects should never be included in global trade or investment of any kind or in the rules that govern trade and investment.

Such complex questions are usually omitted from discussions about global trading systems, which usually keep their focus on new resources, expansion, and profit. But these questions must be addressed if any kind of social or environmental sustainability is to be obtained. How can communal spaces be effectively protected? Do any effective instruments now exist? What new ones can be proposed? How do we define the areas of the commons, common heritage, or government services that should *never* be subject to trade—or at least never be subject to the authority of global agreements that impinge on local or national sovereignty? What are the obligations of nation-states in the modern world? What goods are too dangerous—toxins, weapons, and drugs, for example—to be allowed into the global trade system at all? Should we establish taboos for certain kinds of trade?

CURRENT THREATS TO THE COMMONS

The conversion of the commons into commodified, privatized, "enclosed" form has been under way for centuries. . .In the current context, the main engines of this conversion are the global corporations and the global bureaucracies that increasingly have served these corporations. With the help of the new global trade and finance bureaucracies, corporations are finding opportunities in some virgin territory that most humans never thought could possibly be fodder for corporate enterprise. Here are a few examples.

© Ilin Sergey/Shutterstock.com

Threats to the Freshwater Commons

Unthinkable as it may seem, freshwater—a common heritage basic to the survival of all human beings—is being opened up to private ownership, commodification, export, and trade. It's as if water were an ordinary commodity, like new computer parts or car tires, rather than a shared, irreplaceable, and limited resource needed by all creatures of the earth.

In many parts of the world, the rights to fresh-water in rivers, streams, and lakes are being sold to giant transnational corporations like Bechtel, Vivendi, and others. (Before its collapse, Enron was also a main player in water deals.) These firms have started charging users for every drink of water or liter of irrigation. For people who cannot pay the fees—and many cannot—they, their families, and their fields go thirsty.

Private corporations around the world have identified freshwater as the last great untapped natural resource to be exploited for profit. They arc quickly taking control of water and water services to kick-start trade in what authors Maude Barlow and Tony Clarke have labeled *blue gold*. Indeed, water is becoming just as important as a prior era's black gold—oil.

Companies have been able to proceed aggressively because water has been defined as a tradable commodity by both NAFTA and the WTO. Once the tap is turned on—that is, once any deals are made by *any* state or municipality in a country to privatize water or water services—the tap cannot be turned back off without violating corporate rights. The WTO contains specific provisions prohibiting the use of export controls to prevent the export of water, and NAFTA contains a clause (chapter 11 of NAFTA) that gives companies the right to sue governments for lost *future* profits. This applies, for example, in cases where governments try to stop water export. Water services are also among those slated to be labeled as a commodity in the new General Agreement on Trade in Services (GATS), under a new category called "Environmental Services."

Once water is privatized, commodified, and put on the open market, it is not available to everyone who needs it but only to those who pay. Right now, contrary to popular understanding, most of the world's freshwater is used by corporate industrial agriculture and in manufacturing, such as in the computer industry for manufacture of computer chips. Relatively little is left for drinking or small-scale farming.

Already, the privatization of water has been the subject of enormous angry protests in many countries, notably Bolivia, South Africa, Canada, and elsewhere. In Cochabamba, Bolivia, for example, hundreds of thousands of people rose up against the fees being charged by Bechtel Corporation after it gained control of the municipality's water distribution system and immediately raised prices. On the brink of a revolution, the Bolivian government finally cancelled its agreement with Bechtel, which responded by suing the government under investment rules of a Bolivia-Netherlands trade agreement that mirror the NAFTA rules. As we go to press, the case is still pending.

Threats to the Genetic Commons

Another commons that few people ever thought could be subject to privatization and development is the genetic commons—the vast building blocks of all life on earth. Yet this too is now subject to reinvention through genetic engineering and transformed into patentable commodities.

The late David Brower of Earth Island Institute once called the genetic commons "the last untapped wilderness on earth," but that is no longer true. Like our great forests, the genetic commons are on the verge of rampant commercial intervention. In some areas, like agriculture, the process is well under way. Third World agriculture activists call it *biopiracy*.

According to Andrew Kimbrell of the International Center for Technology Assessment, "Corporations are now scouring the globe seeking valuable plant, animal, and human genes that they can claim as their own private property, as if they invented them. Thousands of gene patents have already been given to corporations, which are now able to patent whole life forms and own them."

Most of this activity falls within the life science industries. Corporations like Monsanto, Novartis, DuPont, Pioneer, and others have benefited enormously from the WTO's TRIPs [trade-related aspects of intellectual property rights] agreement, which confirms their ability to patent plant and seed varieties according to their genetic makeup. Even though these varieties were developed over centuries by indigenous farming communities that shared them with one another freely in a process that is at the core of these cultures, now fees must be paid to use them.

Global corporations insist that this valuable genetic material should not be locked up by small communities but that the whole world should have access to it. Indeed, corporations use the language of the global commons until such time as they confirm their monopoly patents on the material. At that point, all arguments in defense of the commons are abandoned. Instead, the corporations then argue that *they* should be permitted to lock up these genetic materials through patents in order to have a chance of recouping their research investment—for the benefit of all humanity.

Pharmaceutical corporations are especially eager for access and the rights to patent genetic materials. Their representatives travel the globe, exploring traditional native remedies in jungles and fields. They also extract blood and scrape "buccal mucosa" from the skin of native peoples wherever they can, hoping to find genes that contain natural resistance to certain maladies. Usually they accomplish this without disclosing why they are doing it or how much profit they stand to make from their findings and their patents. Among indigenous peoples, the right of "free and prior informed consent" has now become a major international demand before governments can bring in development projects and before companies are permitted to enter.

The cynicism of such practices became especially clear when global pharmaceutical corporations refused to set aside the rules of the WTO's TRIPs agreement in South Africa to permit low-cost, locally developed AIDS drugs to be substituted for the expensive patented varieties they controlled. Only after intense global protest did the patent holders agree to lower prices for AIDS victims there. But the TRIPs rules remain in force for all other instances.

The authors of this document believe that seeds, medicines, and other genetic materials that have been developed in communities for centuries or millennia should always be subject to community control. Any agreement with outsiders to use these materials must be on the basis of fair and equal negotiations on a case-by-case basis, after full discussion of all the relevant facts.

Threats to Communal Lands

In hundreds of cultures around the world, the notion of private, individual land ownership is anathema. Communal land ownership—or no land ownership—is traditional practice and belief among indigenous and farming communities on every continent. This worldview is fundamental to these millions of people, their cultures, their agriculture, and their economic, political, and spiritual practices. In peasant farming communities of South America and Asia and in indigenous communities everywhere, the notion that an individual or corporation could legally gather under its own authority large fruitful tracts of land—thus depriving the people who shared it for millennia—is outrageous beyond all understanding.

Native American activist Winona La Duke has described the land ownership issue among indigenous peoples by pointing out that in her own Oilway culture, the term *nishnabe akin* means "the land to which the people belong." It is a notion that is the exact opposite of Western ideas about land ownership. It is also at the heart of many teachings about humans and land being in *relationship* with each other, with reciprocal obligations.

In contrast, the new trade agreements and policies of international banks and corporations are designed to delegitimize the authority of any such reciprocal arrangements and to solidify private ownership so that land can be more easily bought and exploited. For example, one of the United States' specific demands in NAFTA negotiations with Mexico was that Mexico break up the traditional *ejidos* of the Mayan corn farmers, the system of communal land ownership that began with the successful Zapatista revolution of the early 1900s.

The history of the "enclosure" of the commons is largely devoted to the privatization of communal lands throughout the planet. In most places, the process is so far along that it is rarely publicly argued. However, this has begun to change as mass movements like Brazil's landless peasant movement (Movement of Landless Rural Workers) and other such movements around the world have demanded dramatic land reforms and redistribution under the principle that land, like water, is basic for sustenance.

Appropriation of the Global Commons for Waste Sinks

Ownership and privatization are not the only threats to the commons. There is also the effective appropriation of particular global commons as free dumping grounds and waste sinks for the activity of global corporations.

The atmosphere, oceans, and even outer space have become dangerously polluted, freely appropriated by oil, energy, shipping, and toxic industries as convenient sites to dump effluents and wastes. In the case of automobiles, ships, and the fossil fuel industry, it is an intrinsic result of the technologies that effluents rise to the atmosphere.

When similar pollution takes place in a local commons or inside national borders—say, local smokestack emissions or runoff into rivers—government agencies exist specifically to try to do something about it. This is not to say these agencies do a good job of regulating such activity—they do not—but at least they offer an authority to address the matter and a place where citizens may focus their complaints.

When it comes to the global commons, however, few such agencies exist. There have been some efforts over the last half-century to apply some pollution regulations. The Kyoto Protocol on Global Warming, the Montreal Protocol on Substances That Deplete the Ozone Layer, the United Nations Convention on the Law of the Sea, the Stockholm Convention on Persistent Organic Pollutants agreement, among other multilateral environment agreements (MEAs), are efforts to control impacts on global commons. But all have suffered from tremendous political resistance, weak regulatory regimes, and poor enforcement ability. Still, MEAs remain one hope for rational international recognition, on a case-by-case basis, of the collective rights to maintain the commons in a condition that serves all people and the planet's other species.

An important problem, however, is that when such agreements do begin to make progress in limiting corporate activity, the World Trade Organization can threaten to negate them by asserting its own superior authority on behalf of the primacy of global trade activity. For example, at the November 2001 WTO ministerial meeting in Doha, Qatar, the WTO made explicit its intentions to codify its superiority to MEAs, potentially wiping out generations of effort.

Other solutions to the problem have also been proposed. They include a variety of trust agreements where certain commons would be held as the explicit property of all people in trust for the future. Any encroachments would have to go through very complex permission procedures, thus at least slowing and making visible the problems before they occur.

Sadly, there are hundreds of examples of threats to the commons of the kind we have given here. All reflect increased global pressure on remaining pristine areas. Fortunately, however, there is also increased popular resistance to these encroachments. New methods and instruments must be found or created to prevent continued destruction.

THE TRADITION OF THE COMMONS

In most parts of the world, the tradition of the commons is ages old, though it varies from place to place and culture to culture. These are a few examples.

Europe

In Europe, the concept of the commons dates back at least fifteen hundred years. It referred to commonly shared areas of land and resources enjoyed by all members of village communities, including pasture for the grazing of animals, water from streams and lakes, and all the products of field and forest that people used to sustain their lives. The notion that any of these could be enclosed or turned into the private property of individuals or institutions like corporations was, at first, unthinkable.

The commons deteriorated over time in feudal Europe. There were struggles over definitions of the commons, and large parts of them were appropriated through military force by feudal lords and kings. Still, vast areas remained available for village use as traditional common areas.

As trade developed between regions, and especially with the rise of a large European-wide market for wool, lords and merchants increasingly sought to ensure and expand their own supply by privatizing more and more of the common areas, most importantly land. As they succeeded in enclosing ever larger areas, peasant communities began to lose access to resources they had previously taken for granted. Self-sustainability became more difficult. Rather than remaining as subsistence farmers who also traded in local markets, peasant increasingly became "cottagers"—small artisanal producers—or sought day-labor jobs with the large landowners on lands they had formerly shared. With the coming of the Industrial Revolution, cottage industries declined, and people were entirely removed from their land and sources of sustenance. They became factory workers with no connection to the land. The peasant thus evolved into the proletariat.

Some mainstream economists have justified and praised this conversion of the commons private property as a way of establishing clear ownership and therefore the ability to protect resources from exploitation. But this justification for what was essentially robbery of shared common resources ignores two important facts. First, the people whose lives depended on the commons were almost always excellent caretakers of it and remained so over many centuries. In fact, it was basic to the survival and success of the commons that the entire community shared its values as well as its sustaining virtues and helped protect and preserve it. Second, the people who enclosed the commons, appropriating it for private use, were nearly always outsiders, absentee owners with little personal dedication to nurturing, conserving, or taking care of these resources for the future. The reason they sought private ownership was to exploit the resources as fast as possible. They did not try to own it in order to save it, the assertion of mainstream economists notwithstanding. Rather, they usually pillaged and polluted it.

With the advent of corporate ownership and rights of private property, ownership was made still more abstract or removed from resources, which were increasingly understood only in quantifiable, objective terms. Forests, for example, were no longer appreciated for their community sustainability, the biodiversity they fostered, or their spiritual contexts. Instead, they became "board-feet," ripe for exploitation. That is certainly the situation we face today, in its most amplified form.

Indigenous Communities

In other parts of the world, terms like *commons* were not well-known, but the concepts of shared community use and protection of common resources were basic, endemic, understood and respected by entire societies.

Among indigenous peoples around the world, virtually all political, social, and spiritual values have traditionally been so deeply intertwined with the values and teachings of the natural world that these societies say they are inseparable. It is not really a question of a community commons, as understood by the Europeans. It is more that all creatures—human as well as plant and animal—are directly related, equal, and with equal rights to exist in a fulfilling manner. All economic, political, and spiritual teachings are rooted in that primary relationship.

It is little wonder, therefore, that invading societies—at least those that did not actually slaughter the native populations—made enormous efforts to undermine and destroy their commitment to their traditional relationships to land and nature. That was the only way they could succeed in getting their hands on the resources they desired.

Native peoples were pushed to separate from their lands in hundreds of different ways. Primary among these efforts was the aggressive attempt to undermine traditional religious values and cosmologies as well as traditional native stories and teachings about the need to live in harmony with, and as part of, nature. The actions of missionaries throughout the Americas, the Pacific Islands, and Africa are well-known in this context. They actively helped shift the traditional value system toward a new and more hierarchical view of humans and nature and toward the individualistic notion of private property. In the United States, Australia, and elsewhere, "reeducation" also played an important role—that is, young people were forcibly removed from their traditional communities and placed into boarding schools that did not allow native languages or teachings. The result was a kind of self-loathing that undermined native cultures and their traditional collective economic values.

Equally important were legal maneuvers, such as requiring adjudicatory land title. Since native societies did not traditionally conceive of themselves as "owners" of land but as part of it as a community, "title" and "ownership" of land were an absurdity, but one which they were required to address. The histories of the relationships between invading and native societies are replete with stories of how legalistic maneuvering succeeded in separating Indians from millions of acres of land they formerly enjoyed in a collective manner. These acts of removal and separation continue to this day, notably in the United States.

One particularly appalling recent example is the U.S. Congress's creation of the Alaska Native Claims Settlement Act (ANCSA), purported to be a guarantor of native land rights in Alaska, where these rights had never been abrogated. In fact, ANCSA was the final step in extinguishing the native relationship and rights to enjoy the lands in the traditional manner. Rather than native Alaskans being granted title, or simply granted recognition of "aboriginal rights of ownership" as requested, the Alaskan lands were divided and made into native corporations, managed by native boards of directors, who were given ownership of the lands. But in order to survive, these corporations needed to cut down their forests or exploit their minerals just as any corporation would. These corporate acts, albeit carried out by natives, were contrary to their prior indigenous values. So for the larger society, the desired outcome was achieved; the communal relationship to nature was removed and replaced by an exploitative one, thus providing new fodder for global corporations.

The global invasion of communally held native lands—which has happened on every continent and continues today—has had terrible outcomes, from destruction of the traditional reciprocal relationship between humans and nature to major social breakdowns.

The conflict of values endemic to these pressures over ownership of common lands and resources was very well described in the book *A Basic Call to Consciousness*, published in 1977 by the Iroquois Nation, in a submission to the United Nations Conference on Indigenous Peoples. Here are some short excerpts:

> The majority of the world does not find its roots in Western culture or tradition. The majority of the world finds its roots in the natural world, and it is the natural world, and the tradition of the natural world, which must prevail.
>
> . . .
>
> The original instructions direct that we are to express a great respect, an affection, and gratitude toward all the spirits which create and support life. When people cease to respect and express gratitude for these many things, then all life will be destroyed.
>
> . . .
>
> To this day the territories we still hold are filled with trees, animals, and the other gifts from the Creation. In these places we still receive our nourishment from our Mother Earth. Many thousands of years ago, all the people of the world believed in the same way of life, that of harmony with the universe. [But] the way of life known as Western Civilization is on a death path on which their own culture has no viable answers.
>
> . . .
>
> The Indo-European people who have colonized our lands have shown very little respect for the things that create and support life. We believe that these people ceased their respect for the

> world a long time ago. The air is foul, the waters poisoned, the trees dying, the animals disappearing. Even the systems of weather are changing. Our ancient teachings warned us that if Man interfered with the natural laws, these things would come to be.
>
> . . .
>
> The traditional native people hold the key to the reversal of the processes in Western Civilization Our culture is among the most ancient continuously existing cultures in the world. We are the spiritual guardians of this place. We are here to impart this message.

Asia

Traditional societies everywhere on the planet share values similar to the ones expressed by the North American Iroquois leadership. Concepts like Mother Earth combined with belief in non-hierarchical, nonownership-based communal relationships to field and forest can be found among all peoples who still live in a direct relationship with the earth.

Most people in India today still derive their livelihoods and meet their survival needs from the biological resources of the country, as forest dwellers, farmers, fisherfolk, healers, and livestock owners. Indigenous knowledge systems in medicine, agriculture, and fisheries are the primary basis for meeting their food, health, and cultural needs. In these traditional communities, the biodiversity of the forests and fields and the historic innovations of plant life for food and medicinal purposes have never been seen as the individual property of any person or family but as community resources available to all people. None can be excluded, and neither the state nor any other economically powerful entity can monopolize use of any aspect of the commons, biologically or intellectually.

There is a vibrant struggle in India today over the commons—not only the biological commons (land, forests, water) that have been the basis of sustainability for a great majority of India's population to the present time but also the *intellectual commons*. This refers to the cumulative knowledge that agricultural communities have collected and freely shared for centuries, as well as the innovations they have achieved in developing plant varieties for food and medicine. Global biotech and pharmaceutical companies have been aggressively patenting these examples of the intellectual commons, preventing their common use, and privatizing them for their own purposes. This invasion has led to a level of outrage on the part of India's farmers, indigenous people, and peasant communities that has brought literally millions of people onto the streets in protest against the World Trade Organization's TRIPs agreement, which protects the rights of corporations to engage in these practices.

Community rights to control the biological and intellectual commons are recognized in law as sui generis rights—equivalent to a patent, but recognizing roots in a community rather than an individual. This system is based in part on usufruct rights, which entitle farmers and laborers access to resources needed for their own sustenance, such as common pastures, water, and biodiversity. Sustainability and justice are inherent in such a system of usufruct rights because there are physical limits on how much any one person can labor; hence there are limited returns for labor, unlike capital and private property.

The battle over the biological and intellectual rights of farmers in India, and against the WTO's TRIPs agreement, comes at the end of a long and gainful history of prior enclosures of the commons in India. The policy of deforestation and enclosure of the commons in India began in 1865, when the Indian Forest Act authorized the government to declare forests as "unmeasured" lands or "reserves" for state use. This began what was called the scientific management of forests, but it was really the first step in a long series of moves to remove forests from people and convert them to commodities available for private ownership. Peasant communities that had formerly sustained themselves based on the forests' resources were forced to produce indigo instead of food and pay taxes for salt. They experienced the rapid erosion of their usufruct rights to food, fuel, and livestock pastures, as well as forests and sacred sites. This was a primary cause of their later impoverishment, which led directly to their resistance to further erosion of their common rights by global biotech corporations.

THREATS TO THE "MODERN COMMONS"

In most developed countries, it is hard to remember the time when the central political and economic unit was the local community or when resources were commonly shared. Over the past several centuries, political, economic, and technological evolution in much of the world has conspired to bring far more specialization and industrialization of economic activity, far less economic and social self-reliance, and far greater dependence on dominant centralized political units—cities, states, provinces, and national governments—to provide for the common fundamental needs and services that people require, such as education, transportation, health care, environmental protection, security, and the certainty that there will be sufficient food, housing, and work.

All governments now acknowledge their responsibility in these matters, though they perform them with varying levels of success. Canadians and Danes tend to believe their governments have performed well, at least until recently, whereas Russians and Burmese may not feel this way. Most governments get mixed grades on these matters—good in some ways, bad in others. The United States, for example, has kept its economic and sanitation performance generally high but has been very poor in matters of health care and transportation, and its delivery of these services is marked by extreme inequality. Maude Barlow's forthcoming book, *Profit Is Not the Cure,* details the impact of the World Bank, the IMF, the WTO, and their market-opening policies on the erosion of health care services all over the world, and she chronicles the private sector firms that have benefited. Two decades of these policies have left a legacy of unequal access and health crises for the poor worldwide.

The United Nations Universal Declaration of Human Rights asserts that in addition to the services that nations are expected to perform, governments are also expected to ensure certain basic human rights, religious and political freedoms, and the right to *meaningful* work at fair wages with human dignity. In fact, the U.N. declaration asserts that every person on the planet has an "inherited right to citizenship," which includes health care, education, and work. Moreover, every government has the duty to *defend* the fundamental human rights of every one of its citizens, even beyond its borders. Such assertions also imply that governments have the duty to protect their own ability to perform these services and ensure these fundamental rights against attempts to weaken them.

The authors of this report believe that these services and protections qualify as a kind of "modern commons," in which the state has assumed the responsibility for the common good that once resided in local communities. Even as more power devolves back toward the local—an outcome most of us support—we believe that states have a vital role to play in protecting the community stewardship of both traditional and modern commons. The performance of these responsibilities, however, is now under tremendous threat from global economic institutions.

Great pressure is being applied to national governments by corporations and global bureaucracies to privatize and commodify most public services. Global corporations seek to put them in the same category of commercial activity as toothpaste, cars, real estate services, or movies, provided by private industry at market rates.

Yet there are profound differences between the sacred trust arrangement of governments to provide for basic needs and rights and the private, contracted provision of products, entertainment, and commercial services. Corporations operate from a hierarchy of values that requires profit and growth. Yet as corporations make inroads as providers of health, education, water, or food security, the delivery of these services will go only to people who can pay the market rate. The many who cannot will essentially fall out of the system. *Hence, we would argue that any acts that separate governments from their obligation to provide for all people regardless of economic status are to be prevented.*

Similarly, any efforts by global bureaucracies to *require* that governments privatize such services should be resisted by governments and activists alike—and ultimately banned. At this moment, the current WTO negotiations in Geneva over the expansion of the General Agreement on Trade in Services (GATS) are a case in point.

If the GATS agreement is finalized, then most of the services we have listed here as fundamental rights of citizens and obligations of governments will be subsumed under the new rules of GATS. In other words, corporations will have the right to establish a commercial presence and operate what have until now been domestic services inside countries. These include health care, elder care, child care, water purification and delivery, education, prisons, domestic rail and air transportation, public broadcasting, parks, museums and cultural institutions, social security and welfare programs, and public works of all kinds. As mentioned [earlier], the United States or Canada might find ExxonMobil running public broadcasting while Mitsubishi runs social security; France might have Disney operating the Louvre; Enron or WorldCom could be running the German health care system; and Shell Oil could be in charge of the Japanese railroads and perhaps child rearing as well. Such outcomes are not unrealistic: as we already mentioned, U.S. giant Bechtel was on the verge of running a significant piece of Bolivia's water delivery system until recently, nearly causing a revolution because of its prices to the poor.

The U.N. Covenant on Economic, Social, and Cultural Rights includes the right to education and health. Of all services, these are shaping up as the most potentially lucrative for the global corporations pushing the GATS agreement. Global expenditures on education now exceed \$2 trillion, and global expenditures on health care exceed \$3.5 trillion. Global corporations are aiming at nothing less than the dismantling of public education and health care systems. Already they have succeeded in lobbying over forty countries, including all of Europe, to be listed in the GATS.

The obligation of governments to provide services for their citizens may soon be reduced to whether some distant corporation will or will not charge an affordable rate for people to send their children to school or to a doctor.

In our view, no global agreement should have the power to forcibly intervene in the trust agreement between governments and citizens.

Of course, there are many other threats to basic social programs and services and basic rights to food and health, aside from the upcoming GATS agreement. [Earlier], we discussed the structural adjustment programs of the World Bank and the International Monetary Fund. To be eligible for development loans or debt relief, dozens of developing countries were forced to abandon a multitude of social programs and allow for-profit foreign corporations to enter, commercialize, and privatize these operations to the detriment of local people.

We have also discussed the WTO's TRIPs agreement, which enables global corporations to claim intellectual property rights over the genetic heritage of people and communities and to redefine the role and form of agriculture. Farming and the community way of life it has sustained are becoming dominated by global corporations who have rendered farmers dependent on them for seeds, fertilizers, pesticides, and herbicides. Structural adjustment programs have also played a big role in this shift in agriculture, requiring nations to convert their farms to export-oriented, specialized production and permit entry of global corporations that accelerate the process. This has driven farmers off lands where they formerly grew food for their communities and has increased hunger and migration.

Because food security is one of the most fundamental human rights, every government has an interest in ensuring it as an essential foundation for social stability and public health. We believe this means that no development banks or trade agreements should be given the right to *require* that any country revamp its agricultural production systems, allow foreign entry and investment, or open itself to cheaper food from abroad to its own detriment. Of course, any country may choose to allow such entry if conditions warrant, but it should not be *forced* to do so.

Cultural diversity and integrity is another area that can be seen as part of the commons and is arguably a fundamental right that should be protected by nations. It too is now threatened in many ways by global trade agreements. The right of social, religious, cultural, and indigenous groups to preserve their practices, beliefs, artifacts, and artistic expressions is crucial to maintaining diversity within and among nations and in the world. This also applies to the efforts of nation-states to prevent foreign domination of their own national cultural expressions, through media and artistic creations. Many countries, notably Canada and France—both deeply concerned about retaining film and television industries in domestic hands—have been fighting strenuously to retain their rights to protect their own cultures.

Current WTO trade law subjects culture to the disciplines of the agreement, including "national treatment," "most favored nation," and the prohibition against quantitative restrictions. (All of these WTO rules open up countries to foreign media and other cultural products, often to the detriment of local cultural expressions, which can be overpowered.) There have been several complaints over culture at the WTO since its inception. All have had the effect of limiting the right of a state to protect its cultural industries. The most significant was a 1997 ruling in which the United States successfully forced Canada to abandon protections of its magazine industry, even though American magazines already made up 85 percent of all of those available at Canadian newsstands. Then—U.S. trade representative Charlene Barshefsky said the decision would serve as a useful weapon against Canada's and other countries' protection of their film, books, and broadcasting industries.

The United States is taking such a hard line because any exemption for Canada would set a negative precedent for other countries, especially in the developing world, where cultural protection is an emerging issue. The fact that the effort to undermine these protections failed at the Seattle WTO ministerial meetings does not mean the problems are over. Both the GATS and the TRIPs agreement will have a direct impact on the telecommunications sector, including the Internet, digital and e-commerce spheres, public broadcasting, patents, trademarks, and copyright law. All are now on the table. They should be taken off.

In sum, we argue that in all of the preceding issues, nation-states and communities should not be subjected to the rules of global trade agreements or multilateral institutions like the WTO or the IMF. Trade in these areas should be subject to national and local decision-making processes alone. Only in this way will it be possible for nations to act on behalf of the common interests of their citizens and to fulfill their own obligations to them.

PROPOSALS

At the start of this chapter, we said that the issues raised here—the preservation outside of the global trading system of the remaining nonmonetized, nonprivatized commons and the governmental trust in preserving fundamental services and rights (the modern commons)—would be nuanced and complex. We proceeded from the principle that not every aspect of human experience or nature should be commodified or subjected to the rules of global trade regimes that require adherence to the global free trade model. However, defining exactly which aspects to exclude, and precisely when, is not simple and needs to be the subject of continuing discussion and debate. Stimulating that discussion has been our main purpose in this chapter.

To further the process, we end the chapter with a short list of broad policies consistent with the ten principles for sustainable societies.

The Authority of Trade Agreements Must Be Narrowly Defined Global trade bureaucracies and international financial agencies should not have authority over state or national decision making when it comes to the commons, natural heritage resources, the preservation of national choice in domestic services, or fundamental human rights. Trade and investment agreements should not be allowed to require a national or state government to privatize or commercialize remaining areas of the commons or public services, or to force countries to open up these areas to foreign investment and competition, either by imposing rules requiring governments to conform or penalizing them for not doing so.

Decisions on Common Property Resources and Public Services Should Be Local or National Decisions about common property resources and public services should be reserved to the localities and nations involved, consistent with the principle of subsidiarity. Local commons are the province of local communities. National commons should be subject to national democratic decision-making processes. Global commons should be addressed by multilateral agreements on issues of sustainability and equitable access. No commons—whether local, national, or global—are the proper subject of a multilateral trade agreement. Decisions about the ownership, control, and operation of fundamental public services—including health care and hospitals, water management and delivery, natural resource use, education, transportation, public broadcasting, agriculture and food security, culture, social security, welfare, military, police, and jails—should be specifically excluded from multilateral trade and investment agreements.

Some nations and communities may determine that it is in their public interest to contract for private operation of some of these resources and public services. Indeed, there is sometimes a role for private ownership and markets to play in the management, allocation, and delivery of certain common heritage resources—including land, seeds, and water—but only in a framework of effective, democratically accountable public regulation that guarantees fair pricing, equitable access, quality, and public stewardship. However, decisions in these matters are properly local or national and have no place being prescribed by global trade and investment regimes concerned only with advancing private commercial interests. Nor is it acceptable for any international agreement to dictate actions by local or national jurisdictions that would result in excluding individuals or communities from equitable access to services and resources, such as clean water, that are essential to life and health.

Things Fundamental to Life and, Human Survival Should Not Be Privatized or Monopolized Certain aspects of the commons that are basic to survival should not be privatized or subject to trade agreements. These include the atmosphere, bulk freshwater, and the genetic and molecular building blocks of life (including the human genome). It may be permissible to patent truly distinctive seed varieties created by privately funded research, for a limited period of time, but the patenting of naturally occurring seeds or seed varieties developed by farming communities or by publicly funded research should not be allowed.

Some Aspects of Life Should Not Be Patented or Otherwise Monopolized Here we include some areas of life that *are* now partly privatized and traded but should never be subject to corporate *patent* rights or monopoly ownership as offered by TRIPs. At present, these include genes, seeds, plant varieties, and animal breeds—with the possible exception of temporary rights to the exclusive production and sale of distinctive plant and animal varieties created through privately funded breeding programs. The general principle is this: No patents on life. There is also a need to rethink patent rules on life-sustaining pharmaceuticals—for example, AIDS drugs—to ensure fair pricing and access by all who need them, without regard to financial means.

The Right of Countries to Choose Not to Import or Export Goods They Deem Harmful and Pernicious Should Be Protected; Trade in Certain Pernicious Goods May Properly Be Prohibited by International Agreement A country properly has the right to ban the import or export of certain goods that it considers to present a threat to health and safety, including GMOs, toxins, weapons, and addictive drugs, as well as tobacco and alcohol, for which existing trade agreements now prohibit exclusions. Some currently traded goods are so harmful to the environment, public health, safety, peace, and the global commons that it may be appropriate to create an international agreement to ban them entirely. Candidates would include toxic and nuclear wastes, endangered species, land mines, and sex workers. Such issues, however, are appropriately addressed by international forums and agreements devoted specifically to these topics, not by specialized trade bodies or trade and investment agreements.

International Agreements Are Needed to Protect Global and Transnational Commons Because certain activities in one country can have serious consequences for common resources in other countries, a system is needed to regulate them. For conditions like acid rain, air pollution, ozone depletion, ocean pollution or overexploitation, climate change, and so on, or where common resources such as rivers serve multiple jurisdictions, negotiations are needed on a bilateral or multilateral basis to protect the threatened commons and secure a just and equitable allocation of benefits. These negotiations should take place in forums dedicated to dealing with such issues and should be placed outside the jurisdiction and authority of trade bodies.

Name ______________________ Course ____________ Date ______________

Chapter Questions

1. This article refers to resources known as "the commons". What are some examples of the things traditionally thought to belong to the commons?

2. Which of the current threats to "the commons" discussed in the article do you think is the most serious? Why?

IT'S EASY BEING GREEN

Bill McKibben

In this essay environmentalist Bill McKibben claims that efforts to curb America's demand for energy fall short of what the nation could be doing. Though dated, the essay continues to raise questions of government involvement verses individual initiative in issues surrounding the environment.

The more I surveyed my new car, the happier I got. "New car" is one of those phrases that make Americans unreasonably happy to begin with. And this one—well, it was a particularly shiny metallic blue. Better yet, it was the first Honda Civic hybrid electric sold in the state of Vermont: I'd traded in my old Civic (40 miles to the gallon), and now the little screen behind the steering wheel was telling me that I was getting 50, 51, 52 miles to the gallon. Even better yet, I was doing nothing strange or difficult or conspicuously ecological. If you didn't know there was an electric motor assisting the small gas engine—well, you'd never know. The owner's manual devoted far more space to the air bags and the heating system. It didn't look goofily Jetsonish like Honda's first hybrid, the two-seater Insight introduced in 2000. Instead, it looked like a Civic, the most vanilla car ever produced. "Our goal was to make it look, for lack of a better word, normal," explained Kevin Bynoe, spokesman for American Honda.

And the happier I got, the angrier I got. Because, as the Honda and a raft of other recent developments powerfully proved, energy efficiency, energy conservation, and renewable energy are ready for prime time. No longer the niche province of incredibly noble backyard tinkerers distilling biodiesel from used vegetable oil or building homes from Earth rammed into tires, the equipment and attitudes necessary to radically transform our energy system are now mainstream enough for those of us too lazy or too busy to try anything that seems hard. And yet the switch toward sensible energy still isn't happening. A few weeks before I picked up my car, an overwhelming bipartisan vote in the Senate had rejected calls to increase the mileage of the nation's new car fleet by 2015—to increase it to 36 mpg, not as good as the Civic I'd traded in to buy this hybrid. The administration was pressing ahead with its plan for more drilling and refining. The world was suffering the warmest winter in history as more carbon dioxide pushed global temperatures ever higher. And people were dying in conflicts across wide swaths of the world, the casualties—at least in some measure—of America's insatiable demand for energy.

In other words, the gap between what we could be doing and what we are doing has never been wider. Consider:

- The Honda I was driving was the third hybrid model easily available in this country, following in the tire tracks of the Insight and the Toyota Prius. They take regular gas, they require nothing in the way of special service, and they boast waiting lists. And yet Detroit, despite a decade of massive funding from the Clinton administration, can't sell you one. Instead, after September 11, the automakers launched a massive campaign (zero financing, red, white, and blue ads) to sell existing stock, particularly the gas-sucking SUVs that should by all rights come with their own little Saudi flags on the hood.
- Even greater boosts in efficiency can come when you build or renovate a home. Alex Wilson, editor of *Environmental Building News*, says the average American house may be 20 percent more energy efficient than it was two decades ago, but simple tweaks like better windows and bulkier insulation could save 30 to 50 percent more energy with "very little cost implication." And yet building codes do almost nothing to boost such technologies, and the Bush administration is fighting to roll back efficiency gains for some appliances that Clinton managed to push through. For instance, air-conditioner manufacturers recently won a battle in the Senate to let them get away with making their machines only 20 percent more efficient, not the 30 percent current law demands. The difference in real terms? Sixty new power plants across the country by 2030.
- Or consider electric generation. For a decade or two, environmentalists had their fingers crossed when they talked about renewables. It was hard to imagine most Americans really trading in their grid connection for backyard solar panels with their finicky batteries. But such trade-offs are less necessary by the day. Around the world, wind power is growing more quickly than any other form of energy—Denmark, Germany, Spain, and India all generate big amounts of their power from ultra-modern wind turbines. But in this country, where the never-ending breeze across the High Plains could generate twice as much electricity as the country uses, progress has been extraordinarily slow. (North Dakota, the windiest state in the union, has exactly four turbines.) Wind power is finally beginning

> to get some serious attention from the energy industry, but the technology won't live up to its potential until politicians stop subsidizing fossil fuels and give serious boosts to the alternatives.

And not all those politicians are conservative, either. In Massachusetts, even some true progressives, like the gubernatorial candidate Robert Reich, can't bring themselves to endorse a big wind installation proposed for six miles off Cape Cod. They have lots of arguments, most of which boil down to NIVOMD (Not in View of My Deck), a position particularly incongruous since Cape Cod will sink quickly beneath the Atlantic unless every weapon in the fight against global warming is employed as rapidly as possible.

What really haunts energy experts is the sense that, for the first time since the oil shocks of the early 1970s, the nation could have rallied around the cause of energy conservation and renewable alternatives last fall. In the wake of September 11, they agree, the president could have announced a pair of national goals—capture Osama and free ourselves from the oil addiction that leaves us endlessly vulnerable. "President Bush's failure will haunt me for decades," says Alan Durning, president of North-west Environment Watch. "Bush had a chance to advance, in a single blow, three pressing national priorities: national security, economic recovery, and environmental protection. All the stars were aligned." If only, says Brent Black-welder, president of Friends of the Earth, Bush had set a goal, like JFK and the space program. "We could totally get off oil in three decades." Instead, the president used the crisis to push for drilling in the Arctic National Wildlife Refuge, a present to campaign contributors that would yield a statistically insignificant new supply ten years down the road.

It's not just new technologies that Bush could have pushed, of course. Americans were, at least for a little while, in the mood to do something, to make some sacrifice, to rally around some cause. In the words of Charles Komanoff, a New York energy analyst, "The choice is between love of oil and love of country," and at least "in the initial weeks after September 11, it seemed that Americans were awakening at last to the true cost of their addiction to oil." In an effort to take advantage of that political window, Komanoff published a booklet showing just how simple it would be to cut America's oil use by 5 or 10 percent—not over the years it will take for the new technologies to really kick in, but over the course of a few weeks and with only minor modifications to our way of life.

For instance, he calculated, we could save 7 percent of the gasoline we use simply by eliminating one car trip in fourteen. The little bit of planning required to make sure you visit the grocery store three times a week instead of four would leave us with endlessly more oil than sucking dry the Arctic. Indeed, Americans are so energy-profligate that even minor switches save significant sums—if half the drivers in two-car households switched just a tenth of their travel to their more efficient vehicle, we'd instantly save 1 percent of our oil. Keep the damn Explorer; just leave it in the driveway once a week and drive the Camry.

A similar menu of small changes—cutting back on one airplane trip in seven, turning down the thermostat two degrees, screwing in a few compact fluorescent bulbs—and all of a sudden our endlessly climbing energy usage begins to decline. Impossible? Americans won't do it? Look at California. With the threat of power shortages looming and with some clever incentives provided by government and utilities, Californians last year found an awful lot of small ways to save energy that really added up: 79 percent reported taking some steps, and a third of households managed to cut their electric use by more than 20 percent. Not by becoming a Third World nation (the state's economy continued to grow), not by living in caves, not by suffering—but by turning off the lights when they left the room. In just the first six months of 2001, the Colorado energy guru Amory Lovins pointed out recently, "customers wiped out California's previous five to ten years of demand growth." Now the same companies that were scrambling to build new plants for the Golden State a year ago are backing away from their proposals, spooked by the possibility of an energy glut.

It's only in Washington, in fact, that nobody gets it. If you go to Europe or Asia, you'll find nations increasingly involved in planning for a different energy future: Every industrial country but the United States signed on to the Kyoto agreement at the last international conference on global warming, and some of those nations may actually meet their targets for carbon dioxide reductions. The Dutch consumer demand for green power outstrips even the capacity of their growing wind farms, while the Germans have taken the logical step of raising taxes on carbon-based fuels and eliminating them on renewable sources. Reducing fossil fuel use is an accepted, inevitable part of the political process on the Continent, the same way that "fighting crime" is in this country, and Europeans look with growing disgust at the depth of our addiction—only the events of September 11 saved America from a wave of universal scorn when Bush backed away from the Kyoto pact.

And in state capitols and city halls around this country, local leaders are beginning to act as well. Voters in San Francisco last year overwhelmingly approved an initiative to require municipal purchases of solar and wind power; in Seattle, the mayor's office announced an ambitions plan to meet or beat the Kyoto targets within the confines of the city and four suburbs.

Perhaps such actions might be expected in San Francisco and Seattle. But in June of 2001, the Chicago city government signed a contract with Commonwealth Edison to buy 10 percent of its power from renewables, a figure due to increase to 20 percent in five years. And in Salt Lake City, of all places, Mayor Rocky Anderson announced on the opening day of the Winter Olympics that his city, too, was going to meet the Kyoto standards—already, in fact, crews were at work changing lightbulbs in street lamps and planning new mass transit.

Even many big American corporations have gone much further than the Bush administration. As Alex Wilson, the green building expert, points out, "Corporations are pretty good at looking at the bottom line, which is directly affected by operating costs. They're good with numbers." If you can make your product with half the energy, well, that's just as good as increasing sales—and if you can put a windmill on the cover of your annual report, that's gravy.

In short, what pretty much everyone outside the White House has realized is this: The great economic shift of this century will be away from fossil fuels and toward renewable energy. That shift will happen with or without George W. Bush—there are too many reasons, from environmental to economic to geopolitical necessity, for it not to. But American policy can slow down the transition, perhaps by decades, and that is precisely what the administration would like to see. They have two reasons: One is the enormous debt they owe to the backers of their political careers, those coal and oil and gas guys who dictated large sections of the new energy policy. Those industries want to wring every last penny from their mines, their drill rigs, and their refineries—and if those extra decades mean that the planet's temperature rises a few degrees, well, that's business.

The other reason is just as powerful, though—it's the fear that Americans will blame their leaders if prices for gas go up too quickly. It's not an idle fear—certainly it was shared by Bill Clinton, who did nothing to stem the nation's love affair with SUVs, and by Al Gore, who, during his presidential campaign, demanded that the Strategic Petroleum Reserve be opened to drive down prices at the pump. But that's what makes Bush's post-September silence on this issue so sad. For once a U.S. president had the chance to turn it all around—to say that this was a sacrifice we needed to make and one that any patriot would support. It's tragically likely he will have the same opportunity again in the years ahead, and tragically unlikely that he will take it.

In the meantime, there's work to be done in statehouses and city halls. And at the car lot—at least the ones with the Honda and Toyota signs out front. "This Civic has a slightly different front end and a roof-mounted antenna," says Honda's Bynoe. "But other than that, it looks like a regular Civic, and it drives like one too. It's not necessarily for hard-core enviros. You don't have to scream about it at the top of your lungs. It's just a car." But a very shiny blue. And I just came back from a trip to Boston: 59 miles to the gallon.

Name ______________________ Course ____________ Date ______________

Chapter Questions

1. How easy would it actually be for Americans to free themselves from their national oil addiction? In your opinion, why do most Americans choose not to reduce their dependence on oil?

2. Do you think that hybrid cars would ever take over the majority of American vehicles? Why or why not?

3. Of the numerous suggestions that MacKibben argues for, which seem the most conceivable for people in modern society?

SECTION 3 Communication: The Transfer of Information over Space and Time

© jorgen mcleman/Shutterstock.com

Introduction

Thousands of years ago, when all humans lived as hunter-gathering tribes, oral tradition was satisfactory for passing information among them. It relied on things we most easily remember: stories, and songs. But with agriculture and settled existence came complex societies with large populations that had to be managed and supplied. They relied on items that people do not easily recall like numbers and lists. The answer to this problem first appeared in Mesopotamian city-states around 3000 BC. Shapes pressed into wet clay with a reed stylus represented things, ideas, or numbers and were preserved when the clay hardened. Writing, our first communication technology had appeared. But, this "Cuneiform" writing contained hundreds of symbols which kept it an esoteric knowledge for a few scribes in each city.

Though simpler, alphabetic systems later appeared, writing remained the possession of the Powerful. In medieval Europe, kings and the Church controlled the spread of ideas through control of writing, as most writers and copyists were in their employ. That changed in the 1440s when a German metalsmith named Johann Guttenberg fashioned a printing press, the second major communication technology after writing. Although he did not invent block printing, Guttenberg's invention made it a mass production process that ushered in the Age of the Book and dramatically quickened the spread of ideas.

With the development of the telegraph in the 1830s and 1840s, electricity became, in one way or another, the basis for most communication innovations. By 1876, Alexander Graham Bell found ways to transfer the human voice into an electric signal that could travel the telegraph wires. A speaker at the other end created sound waves that simulated the caller's voice. A year later, Thomas Edison's phonograph introduced a way to convert sound waves onto a cylinder and play it back with an amplifier that mimicked the original sound.

The telephone and phonograph reproduced sound as a type of False Reality. This occurs when the original sound waves are converted into another medium, such as an electric current, an analog wave, or, more recently, a digital signal. It is then stored or transferred in this form till it is converted once again into sound waves by a loudspeaker. By contrast, Immediate Reality occurs when the producer of the sound and the sound is not changed or amplified. False Reality allowed users of the telephone to hear messages in person rather than reading a clipped telegraph message. Bell, who originally imagined his invention primarily as an aid for the deaf, found it rapidly spreading across the country.

Edison followed his phonograph with a functional light bulb, and, then, in the early 1890s, told lab assistant William Dickson to combine those inventions in a machine that made use of the brain's ability to interpret images flashing at least sixteen times per second as movement. Using continuous plastic strips to store exposed photographs in succession, Dickson's Kinetoscope introduced the world to film and to a visual element of False Reality. Early filmgoers sometimes failed to recognize the line between False and Immediate Reality. For example, when France's Lumiere Brothers, inventors of a large screen projector, showed simple footage of a train arriving at a station on the wall of converted theaters, many patrons scrambled for the exits believing a real train was crashing through the wall. The thrill of False Reality increased in the late 1920s with the introduction of sound film technologies, which caused film's popularity to soar. By 1930 sixty-five percent of the American population visited the theater at least once each week.

This print captures the earliest surviving copyrighted motion picture, **Edison Kinetoscopic Record of a Sneeze**, that was made by W. K. L. Dickson in January 1894.

Another, seemingly more miraculous, form of communication kept pace with film in this era. Scientists had tinkered with electromagnetic waves in laboratories since the late 1800s but never produced a practical application. It was left to an Italian "tinkerer" named Gugliemo Marconi to build a practical and commercially viable method of wireless communication. Marconi's early radio acted as a wireless telegraph, though other inventors soon found ways to send radio signals that caused speakers in the receiver to mimic voices and music. The ability of radio waves to travel hundreds of miles instantly meant that communication had now defeated space and time.

The pinnacle of communication technologies seemed to have been reached by the 1930s with the invention of the television. This was really a host of innovations that produced electromagnetic waves, which, in a receiver set, created and flashed pictures in sequences that simulated motion. Film and radio had been joined, and television broadcasts quickly spread over much of the developed world.

The last communication revolution, the Internet, which some claim as the most important of them all, combined every innovation before it and dramatically increased the amount of information that could be moved. Secret experiments commissioned by the U.S. military in the 1960s resulted in the ability of digital information stored in computers to traverse phone lines. By the time the Internet came into widespread public use in the 1990s, it was becoming clear that the mass of information generated by computers required larger and faster mediums than the phone lines, whose capacity had been little increased since the first ones were strung a century earlier. Cable, fiber-optic lines, and most significantly, electromagnetic waves, have been called into service over the last thirty years and made the Internet the fastest, most far-reaching advancement in humankind's efforts to transmit information.

Communication, which first was advanced by writing and the printing press, may have found its ultimate vehicle in the Internet, which, today, threatens to invade even the Earth's most remote locations. However, as this brief history has indicated, there is little reason to believe it will be the last major form of communication that tremendously affects society.

THE AXEMAKER'S GIFT

James Burke and Robert Ornstein

In The Axemaker's Gift, Burke and Ornstein explore the effects of history's major innovations. In this excerpt, they theorize that society's elites used writing to "cut," or manage the ordinary citizen, but warn that this gave the elites unprecedented "control."

© Kamira/Shutterstock.com

Tablet with Cuneiform writing

The new gift of writing by this patterned process was still, however, extremely complex and esoteric, consisting as it did of at least 2,000 different signs. There were, for instance, no fewer than thirty-one versions of the sign for "sheep" in various states, conditions, and types. Mastery of these signs required many years of training and made the art of writing a highly specialized skill known only to a few. The inaccuracy of the phonetic renderings of signs (representing in many cases closely similar sounds) as well as the confusion created (because the same sign often could be read either as the picture of a sound or of an object) all made the learning process lengthy and difficult. While writing may well have facilitated the development of a much more complex and heterogeneous community, giving many of its members a style of life undreamed of only a few generations earlier, the rewards came at a cost. The power of writing to organize and command was still inaccessible to all but the very few.

The scribal bureaucracies maintained control of taxation and the allocation of resources, of workers' remuneration, and of domestic and foreign trade. Later, in Egypt, the representations of economic activities on monuments would invariably show a scribe in a prominent location,

---

keeping track of the operation. Because of their importance in the administration, the scribes held a respected and privileged position, so they had no interest in simplifying their arcane art, since that would serve to make it more accessible to potential competitors.

Administration of such a system required a large, specialized, professional clerical establishment. The scribal school, or *edubba*, was the driving force behind society and generated a small, literate elite. The schools came into existence early in the third millennium, and the full-time course lasted from childhood to early adulthood. Study began with practice of the syllabic characters, such as *tu ta ti*, *nu na ni*, *bu ba bi*, etc., after which a repertoire of some 900 signs was taught, and this was followed by groups of more than one character. Following memorization and practice of these fundamentals came the real reason why it was difficult for many people to share the knowledge of reading and writing. The student was confronted with months of study in order to learn lists of thousands of items arranged by subject matter, like parts of animals and the human body, names of domestic animals, of birds and fish, plants, implements, etc. The difficulty with pictograms was that there were almost as many sign patterns as there were things to be signed.

In their literary, linguistic, mathematical, and astronomical achievements, the scribal schools far exceeded practical and bureaucratic needs, and in doing so (small wonder) widened the division between the leaders and the led. The schools, new arenas of intellectual activity created by axemaker technology and closed to the majority of the population, used literacy to develop an unprecedentedly comprehensive, though extremely selective, educational system, and through it they established a powerful elite. Many of the thousands of surviving tablets give the names of the scribes and even the names and occupations of their fathers. As might be expected in a society where literacy was a jealously guarded skill, scribes came from important and wealthy families and many went on to high office in the administrative bureaucracies.

Then, in a process that would be repeated again and again throughout history, the qualified few were obliged by changing circumstances to diffuse some of their specialized skills. As technology began to generate increasing social change, the unavoidable alternative to social collapse was to allow a wider segment of the community to read and write. The number of pictographic signs was dramatically reduced, over only five centuries, from two thousand to three hundred, and, in consequence, became much more generally used.

Even though the proportion of the population able to use the new, simpler pictographs was still less than one percent, it was now possible to organize a much more complex social structure and to institute the beginnings of a properly bureaucratic cutting-up of social activity. Several cities joined in federations finally to be ruled by a single king. With this arrangement came the regularization of a wider view of the "place" as a complex, rigidly layered, hierarchical entity, which we in the modern world describe as the "state." The apparatus of community organization was now established in at least three classes: lowest-ranking administrators (overseers and foremen, supervising laborers and farmers), the overseers of the overseers (working in offices, running schedules), and, at the highest level, the policymakers.

This division of labor was made possible by the technology of writing and, for the same reason, so too was conformity in public thought and behavior on a scale not possible before. In simpler times, orally transmitted rituals and orders might, however well-remembered and rehearsed, have become deliberately or unconsciously distorted, but now they could be made permanent and codified in writing. This left little room for the deft avoidance of duty. Axemaker writing unified the Mesopotamian command-and-control system through the bureaucracy, and by the third millennium this authority began to extend beyond work-oriented organization into the behavior of individuals in their private lives.

What happened next was a phenomenon that singles out Mesopotamia among the other great riverine civilizations of the period in India, China, and Egypt. In Mesopotamia, the extension of social control through the use of literacy radically altered the relationship between individuals and between them and the city authority, thanks to the invention of law.

One of the earliest examples of the new rule of law was the way it took redress of grievance out of the hands of a victim's kin. The way it did so was special to Mesopotamia. Earlier, in hunter-gatherer times, it seems that wrong-doing had been punished through acts of revenge by family members, in which the injured parry or their kin acted as both judge and executioner. The problem with this system was that revenge could turn the original wrongdoer into an injured party, whose kin might then in turn desire their own revenge. This process could all too easily give rise to blood feuds lasting for generations. This kind of "loose cannon" behavior within the close confines of city walls could easily damage the cohesion of the community. The creation, by axemaker tools, of a "place," in the form of these first tightly organized cities, had now begun to affect the values and ethics of those living a highly regimented life within their walls.

In India, China, and Egypt, as religion became institutionalized, the responsibility for judgment and punishment seems to have been transferred from heads of families to the overall authority of the priesthood. Antisocial acts were made even more undesirable by being defined as an affront to the gods.

However, Mesopotamia broke the pattern, thanks to an early shift from communal to private ownership of property as far back as 4,500 years ago, when Mesopotamian clay tablets were already carrying details of private transactions and contractual agreements. Perhaps this kind of low-level commercial interaction needed something simpler than the full ceremony of reference upwards to the gods, and for that reason a lower-level form of regulatory authority came into being.

This, the first written law, differed from other institutions set up so far for the purpose of social control. Since it dealt primarily with private property, it also had to take into account a radically new concept. The axemakers had provided a gift that, throughout history from then on, would color all social development—the idea of the individual's rights and duties of ownership. For the first time, a template for behavior existed that could be adapted to any individual circumstance. Individuals' lives were no longer to be lived entirely at the whim of priest or king. On the other hand, those lives were no longer entirely their own, thanks to the development of law.

All early Mesopotamian law is little more than a collection of precedents, key among which are the references to the special form of the ruler's authority. This authorized the intrusion of power into the private lives of each individual on pain of heavenly retribution. All edicts begin with a statement of the king's appointment by the gods to rule over the city-state, then follows the body of law, and the edict ends with a warning that anybody defying the king's law will be cursed.

Four thousand years ago, in what may be a reference to the first attempt at rule by law, King Ur-Engur of the city of Ur says he administered justice "following the laws of the gods." This gave added magic force to the new code for behavior by suggesting that any transgressor who disobeyed the king would be punished by authorities as far away as heaven. You could run, but you couldn't hide.

The oldest known formal legal code is that of King Ur-Nammu of the city of Nippur, founder of the third Ur dynasty, 4,050 years ago. Among other things, it refers to another major element of social control, reflecting the extent to which the axemaker gift of numbers had developed to become a tool to standardize behavior and regulate human relationships in commerce. This new constraint on thinking came in the form of a standardized system of weights and measures. The world could now be officially packaged. Increasingly, standardized commodities and standardized behavior life was becoming routine.

The relevant court of law imposing order on disputes was, not surprisingly, the temple. Hundreds of tablets unearthed in the ruins of Nippur, in the area where the temple scribes lived, all paint a clear picture of the administration of justice during the period. Court records were known as *ditilla*, meaning a "completed lawsuit," and among them are notarizations of agreements as well as contracts referring to marriages, divorces, child support, gifts, sales, inheritance, slaves, and the hiring of boats, pretrial agreements, subpoenas, thefts, and damage to property. Justice was administered by the *ensi*, the governors of the cities within the city-state, who represented the king. The temple served as the courtroom and there were no professional judges, since the thirty-six men listed as judges were from all walks of life: merchants, scribes, temple administrators, and high officials of some kind.

From the beginning of the third millennium B.C.E., five thousand years ago, groups of *Amurru*, Amorite Semitic invaders from the Syrian and Arabian deserts, made sporadic but increasingly frequent attacks on Mesopotamia. The city-states finally fell, the Sumerians ceased to exist as an ethnic, linguistic, and political entity, and the Amorites established a dynasty that was to rule from its center at Babylon for three hundred years. In 1792 B.C.E. the Babylonian king Hammurabi came to power, and under his supreme authority Babylonian government became highly centralized, with a tight network of governors and officials, who represented the king's interests in all aspects of public life.

Forty-two years later Hammurabi issued the now-famous legal Code of Hammurabi. In its prologue, he claims to have been elected by the gods to reign over Babylon and to preserve justice among its people, ending with blessings for those who respect its laws and curses for those who do not. The Code's epilogue states: "These are the prescriptions for justice established by Hammurabi, in accordance with the wishes of the gods, who guided the kingdom along the proper [*sic*] path." The epilogue reveals an awareness of the strength of legislation as a tool of social reform, designed to prevent oppression and facilitate justice.

The Code was divided into three parts, the central section containing 282 clauses and showing clear evidence of a drive toward more permanent, systematic legislation. Significantly, the Code is carved on a durable stone column, not written on fragile tablets of clay. The provisions of the code are more numerous and much more secular than the codes that preceded it. And for the first time, the "eye for an eye" canon is introduced as a deterrent. The law, for instance, now mandated the death penalty on occasions where, before, it might have been enough to require retribution payment, sanctioned by the temple and due from the wrongdoer's kin.

The Mesopotamian concern for the continuity of social order, ensured by the rigid codification of the individual's rights and responsibilities in law, has informed the whole of subsequent Western thinking, especially in the way it supports the separation of society into classes under the supreme monarch, who rules by divine right. A major example of the way axemaker gifts profoundly modify our thinking lies in the way this gift (that so drastically reduced our ancient, hunter-gatherer liberties) radically altered our view of individual behavior, to an extent that still colors our attitudes thousands of years later. We in the modern world refer to "freedom under law" when we speak of what our ancient ancestors would surely have regarded as major constraints on all their freedoms.

But the modification of thinking brought about by writing and commerce pales into insignificance beside what happened because of the interaction between them. Trade in each of the riverine civilizations was facilitated immensely by the esoteric ability to write, and it was in Egypt that key advances in writing technology occurred, thanks to the plentiful supply of a much more versatile and portable writing medium than the clay tablets of Mesopotamia. Although the presence of the state authority was always visible in the form of hieroglyphs on virtually every structure, like Soviet factory slogans, reminding the people of the plans and achievements of omnipotent Pharaoh, the Egyptian economy was driven by a plant. The papyrus, growing in profusion along the Nile, made possible the easy spread of literacy among bureaucrats because its leaves were prepared simply and written on with brush and ink.

Two versions of writing developed, one called "hieratic," for use in religious writing and official statements, and the other a simpler, popular form, called "demotic," that dealt with abstract concepts. With a highly flexible form of communication like this, the Egyptian elite soon built an empire unrivaled in the Mediterranean. Egyptians traded with China through Indian middlemen; they went to the Atlantic and traveled as far south as central Africa.

Egypt developed slightly later than Mesopotamia and, given her differing environmental circumstances, took an alternate route to civilization. To begin with, the hunter-gatherers in Egypt had long been protected by natural barriers against enemy incursion: to the north lay the sea, and in all other directions, the desert. There was no need, as there had been in Mesopotamia, to develop separate, independent village-cities, each self-protected from attack. The flooding of the Nile was extremely regular, making the early development of large-scale, centralized public-works irrigation schemes relatively easy.

From the beginning, in Egypt, the common experience of a great life-giving river may have originally generated common myths and beliefs that were shared by tribal communities separated by distance but united by the Nile and making their later integration easier. Given the homogeneous nature of the community, it was relatively easy for a single, supreme authority to appear almost from the start, and by 5,000 years ago the centralization of authority in Egypt was total.

Pharaoh's administrators were members of his family, he ruled alone by divine right, and there was no legal system like that in Mesopotamia, since there is no mention of merchants and the concept of private property that had generated Mesopotamian legal codes. Egyptian bureaucracy was closely associated with the temple-court, and it controlled all trade in and out of the country. Since everything in Egypt literally belonged to the Pharaoh, the distributor of all benefits, he alone issued governmental edicts regarding everything from regulations for timber cutting to irrigation, ship-building, agricultural practices, and commercial travel.

Egypt attained the most bureaucratized economy in history, thanks to the rigid stratification of its community. The extreme, centrally organized division of labor generated an economy that consisted basically of specialized, exclusive crafts and massive state-controlled labor projects, like irrigation canal networks, sacred cities of the dead, and pyramids. What the Egyptians accomplished was not very different or innovative, but it was uncommonly big.

The dynastic Egyptian state marked a new stage in the expression of power. The institutionalization of control through technology and writing enshrined preferential treatment for the literate. The gulf between the aristocratic elite and the powerless, passive commoners was sanctioned by practice and ritual. In light of this, it is not surprising that here, as in all ancient communities, the first law to be formalized by the central authority regarded acts of *lèse majesté*.

Almost all of the technological advance was pressed into service to help government to function and to control the community. Writing and mathematics serviced taxation and organization. Specialized metallurgical crafts created weapons of conquest and luxury objects of veneration. Calendrical knowledge, astronomy, and geometry were developed specifically for state irrigation projects or to invest the authorities with the magic power of eclipse prediction.

And then, approximately 3,600 years ago, came a major development that would make the acquisition and application of knowledge immensely easier and that would once again alter the way the Western brain worked. Its emergence would also signal the beginning of the end of our millennial reliance on tradition, ritual, and divine authority. The new axemaker product would free rulers from any limitations on their freedom of action that oral tradition might have imposed, because it made the management of higher rates of innovation and change much easier. It was a new kind of script, one that was the first truly transparent communications system because it could be used to express any and all languages. It was the alphabet.

It first appears in one of Egypt's overseas enterprises, a turquoise mine in the mountains of the southern Sinai peninsula, at a place now called Serabit el Khadem. The building complex there included a temple to Hathor, goddess of turquoise, and a large building, including a courtyard, sanctuaries, baths, and soldiers' barracks. The mining personnel were Semitic slave laborers and the enterprise was managed by Canaanites, who spoke a Semitic language close to ancient Hebrew.

These Canaanite mining specialists had been trained at an Egyptian mercantile center and would have been aware of the main commercial forms of writing at the time—hieroglyphs and pictographs—neither of which suited the Canaanite language and both of which were still difficult and complex to write. It might have been the search for an easier way to do things that drove one of the Canaanites to think up a simpler form of expression. Or perhaps the invention had happened elsewhere among the Semites and was brought to Sinai by the miners. But whoever did it and for whatever purpose the new tool was invented it must have made intercommunity trade and technological advance easier almost at a stroke.

The Egyptians (and the Mesopotamians, Cretans, Cypriotes, and Western Semitic peoples) had already begun to abbreviate their complex forms of picture-writing by using syllabaries. A syllabary reduced the number of signs by adopting a common sign for all examples of a single consonant. This made much easier the use of Egyptian hieroglyphs, for instance, that included no fewer than 700 signs. The syllabary was achieved by taking the sign for a word with that consonant's sound and making it the sign for the sound wherever it appeared (e.g., "mayem," the wavy sign for "water," was used to represent the letter "M" because the spoken word had an "M" sound).

However, what made the syllabary difficult to use easily in other languages (unlike the alphabet) was the way that, in Egyptian, for instance, twenty-four signs were used to express all the possible modifications of a consonant by a vowel ("ma," "mo," "mi," etc.), and a further eighty signs represented a pair of consonants modified by *two* vowels (e.g., "timi," "tama," "tima," etc.). And since some of the vocalic sounds involved in syllabaries were not necessarily simple vowels, or they were present only in one particular language, in a kind of linguistic catch 22, it was necessary for a reader to know the language in order to reproduce the sound.

The Sinai scribe probably developed his technique from a syllabary known to him (perhaps Western Semitic, used in Phoenician, Hebrew, and Aramaic), by simplifying and reducing the number of letters needed. All he did was remove the modified forms. In this way, a written transcription of the sound system could satisfy the needs of any language, without having to refer to the full pictographic signs that went together to make up the word.

The signs found at Serabit el Khadem, scratched on the limestone rock faces, are of letters written in a flowing style, indicating an origin in brush and ink rather than in carved stone. The unknown Canaanite axemaker had invented the first true alphabet in order to make it easier to do business deals between members of different linguistic groups, but when the new script eventually reached Greece, it would do much more: it would trigger the beginning of modern thought.

So in the near ten-thousand-year period, from the first agricultural settlements to the creation of numerals and the alphabet community, hierarchs had used the axemaker gifts to maintain, strengthen, and centralize their grip on society while at the same time providing an increasing number of its members with the means to lead fuller and more satisfying material lives. But the gulf was widening all the time between the few who understood the esoteric knowledge that conferred the power of social cut-and-control and the many who did not. And in spite of the fact that, from the shaman's baton to the alphabet, toolmaking generated an ever-growing amount of ever more accessible knowledge, it must be remembered that at no time was this access ever available to more than a tiny fraction of the population.

And as knowledge expanded, so did specialization and esoteric practices. Above all, increased knowledge created more complex societies and activities, and these demanded ever more careful management. The consequences of social breakdown, in the crowded confines of a Mesopotamian city reliant on organizational conformity to ensure the continued provision of food, were potentially much more dangerous than they would have been among a small free-ranging, hunter-gatherer group ten thousand years earlier. Besides, the prospect of security and continuity must have been attractive not only to the few who would profit immensely from a concentration of labor inside city walls, but also to the great unwashed, living on the precarious line between feast and famine, hemmed in by the same walls and cut off from the source of their food and clothing. Conformity and obedience in these circumstances was pragmatic. The devil, you say.

Even by this early date, axemaker gifts had already given us the ability to perform miracles. We had used them to emerge from the jungle, first to small, regularly fed agricultural settlements and then to large, well-ordered cities. There, in return for the security of protection, possessions, and food, we traded the ancient hunter-gatherer freedom of movement and the right to change our leaders, for royal dynasties that ruled by divine right and codified our behavior with the rule of law.

Concentrated and regimented in the cities, bound by rigid conformity, we were conveniently ready for the next great axemaker change. In return for the gift of the alphabet we had accepted, we would have to accept a new measure of conformity in the way we thought about thinking.

Chapter 2 Token Contribution

Adams, Barbara. *Predynastic Egypt*. Princes Risborough: Shire, 1988.

Alizadeh, Abbas. "Socio-Economic Complexity in Southwestern Iran During the Fifth and Fourth Millennia B.C.: The Evidence from Tall-i Bakun A." *Iran* 24 (1988).

Astle, Thomas. *The Origin and Progress of Writing*. 2d ed. London: J. White, 1803.

Cherry, Colin. *On Human Communication*. London: John Wiley and Sons, 1957.

Chiera, Edward. *They Wrote on Clay*. Chicago: University of Chicago Press, 1938.

Childe, G. *Man Makes Himself*. Harmondsworth: Penguin, 1936.

Claessen, Henri J. M., and Peter Skalnik. *The Early State*. The Hague: Mouton, 1978.

Clark, Grahame. *World Prehistory: A New Outline*. Cambridge: Cambridge University Press, 1969.

Diakonoff, Igor M. "Some Reflections on Numerals in Sumerian Towards a History of Mathematical Speculations." *Journal of the American Oriental Society* 103, no. 1, (1983).

Driver, G. R. *Assyrian Laws*. Oxford: Clarendon Press, 1935.

Driver, G. R., and J. C., Miles, The Babylonian Laws Vols I & II. Oxford: Clatendon Press, 1952.

Finlay, F. *Ancient Slavery and Modern Ideology*. Harmondsworth: Penguin, 1983.

Frankfort, Henri. *The Birth of Civilization in the Near East*. London: Williams and Norgate, 1951.

Fry, Edmund. *Pantagraphia*. London: Cooper and Wilson, 1977.

Haas, Jonathan. *The Evolution of the Prehistoric State.* New York: Columbia University Press, 1982.

Henry, Donald O. *From Foraging to Agriculture.* Philadelphia: University of Pennsylvania Press, 1989.

Innis, Harold A. *Empire and Communications.* Oxford: Clarendon Press, 1950.

Jackson, Donald. *The Story of Writing.* New York: Taplinger Publishing Company, 1981.

Jasim, Sabah Aboud, and Joan Oates. "Early Tokens and Tablets in Mesopotamia: New Information from Tell Abada and Tell Brak." *World Archaeology* 17, no. 3, (1986).

Kramer, S. N. *The Sumerians.* Chicago: University of Chicago Press, 1963.

Kraus, R. C. *Brushes with Power.* Berkeley: University of California Press, 1991.

Kutas, M., S. A. Hillyard, and M. S. Gazzaniga. "Processing of Semantic Anomaly by Right and Left Hemispheres of Commissurotomy Patients: Evidence from Event related Brain Potentials." *Brain* 111 (June 1988).

Lieberman, S. J. "Of Clay Pebbles, Hollow Clay Balls, and Writing: A Sumerian View." *American Journal of Archaeology* 84, no. 3, (1980).

MacDowel, Douglas M. *The Law in Classical Athens.* London: Thames & Hudson, 1978.

Macneill, William. *The Rise of the West.* Chicago: University of Chicago Press, 1963.

Maisels, Charles Keith. *The Emergence of Civilization: From Hunting and Gathering to Agriculture, Cities, and the State in the Near East.* London: Routledge & Kegan Paul, 1990.

Menninger, Karl. *Number Words and Number Symbols.* Cambridge: MIT Press, 1977.

Mote, Frederick W. *The Intellectual Foundations of China.* New York: Alfred A. Knopf. 1989.

Naquin, Susan, and E. S. Rawski. *Chinese Society in the Eighteenth Century.* New Haven: Yale University Press, 1987.

Neufeld, E. *The Hittite Laws.* London: Luzac & Co., 1951.

Pfeiffer, J. E. *The Emergence of Society: a Prehistory of the Establishment.* New York: Harper & Row, 1977.

Robson, William A. *Civilisation and the Growth of Law.* London: Macmillan, 1935.

Sampson, Geoffrey. *Writing Systems.* Stanford: Stanford University Press, 1985.

Schmandt-Besserat, Denise. "From Tokens to Tablets: A Re-evaluation of the So-called Numerical Tablets." *Visible Language* 15, no. 4, (1981).

———. "The Emergence of Recording." *American Anthropologist* 84, no. 4, (1982).

———. "The Envelopes that Bear the First Writing." *Technology and Culture* 21, no. 3, (1980).

———. "Tokens at Uruk." Cat. nos. 23, 104, and 107; Susa Cat. nos. 170, 658, and 659, BWII, (1988).

———. "The Use of Clay Before Pottery in the Zagros." *Expedition* 16, no. 2, (1974).

———. *Before Writing:* Volume 1, *From Counting to Cuneiform.* Austin: University of Texas Press, 1992.

Service, Elman R. *Origins of the State and Civilization: The Process of Cultural Evolution.* New York: Norton, 1975.

Slobin, D. I. "Universals of Grammatical Development in Children." *Advances in Psycholinguistics,* edited by Flores d'Arcais, G. B., and W. J. M. Levelt. Amsterdam: North-Holland Publishing, 1970.

Smith, Philip E. L. "Ganj Dereh Tepe." *Paléorient* 2, no. 1, (1974).

Solecki, Ralph S. *Shanidar.* Harmondsworth: Penguin, 1972.

Trigger, B. G., et al. *Ancient Eygpt: A Social History.* Cambridge: Cambridge University Press, 1983.

Vallat, François R. *Le Natoufien.* Paris: Cahiers de la Revue Biblique, no. 15 (1975).

Wright, Henry T. "An Early Town on the Deh Luran Plain, Excavations at Tepe Farukhabad." *Memoirs of the Museum of Anthropology* 13. Ann Arbor: University of Michigan, 1981.

Name ______________________ Course ______________ Date ________________

Chapter Questions

1. The Cut and Control theory, presented in this selection, implies that, though identification numbers like social security numbers and driver's license numbers have a function, at the same time, they allow the elites too much control over other members of society. Do you agree that the potential for harm outweighs the organizational good of these numbers?

2. Explain the phrase "esoteric knowledge" as it is used in this selection.

THE PENCIL

Henry Petroski

Some objects are so obvious they are forgettable. American engineer Henry Petroski believes the pencil is one such object. In this 1989 writing, the author traces the origins of this tool and its many uses in history. He mentions that naturalist and author Henry David Thoreau, whose family was in the pencil making business in the mid 1800's, failed to list the pencil on his list of essential items for his 12 day excursion into the Maine woods. Petroski writes of Thoreau's omission, "Perhaps the very object with which he may have been drafting his list was too close to him, too familiar a part of his own everyday outfit, too integral a part of his livelihood, too common a thing for him to think to mention." This story helps readers realize the usefulness of certain items often taken for granted, often hiding in plain sight.

Henry David Thoreau seemed to think of everything when he made a list of essential supplies for a twelve-day excursion into the Maine woods. He included pins, needles, and thread among the items to be carried in an India-rubber knapsack, and he even gave the dimensions of an ample tent: "six by seven feet, and four feet high in the middle, will do." He wanted to be doubly sure to be able to start a fire and to wash up, and so he listed: "matches (some also in a small vial in the waist-coat pocket); soap, two pieces." He specified the number of old newspapers (three or four, presumably to be used for cleaning chores), the length of strong cord (twenty feet), the size of his blanket (seven feet long), and the amount of "soft hardbread" (twenty-eight pounds!). He even noted something to leave behind: "A gun is not worth the carriage, unless you go as a huntsman."

Excerpt from *The Pencil: A History of Design and Circumstance* by Henry Petroski, 1989.

Thoreau actually was a huntsman of sorts, but the insects and botanical specimens that he hunted could be taken without a gun and could be brought back in the knapsack. Thoreau also went into the woods as an observer. He observed the big and the little, and he advised like-minded observers to carry a small spyglass for birds and a pocket microscope for smaller objects. And to capture the true dimensions of those objects that might be too big to be brought back, Thoreau advised carrying a tape measure. The inveterate measurer, note taker, and list maker also reminded other travelers to take paper and stamps, to mail letters back to civilization.

But there is one object that Thoreau neglected to mention, one that he most certainly carried himself. For without this object Thoreau could not have sketched either the fleeting fauna he would not shoot or the larger flora he could not uproot. Without it he could not label his blotting paper pressing leaves or his insect boxes holding beetles; without it he could not record the measurements he made; without it he could not write home on the paper he brought; without it he could not make his list. Without a pencil Thoreau would have been lost in the Maine woods.

According to his friend Ralph Waldo Emerson, Thoreau seems always to have carried, "in his pocket, his diary and pencil." So why did Thoreau— who had worked with his father to produce the very best lead pencils manufactured in America in the 1840s—neglect to list even one among the essential things to take on an excursion? Perhaps the very object with which he may have been drafting his list was too close to him, too familiar a part of his own everyday outfit, too integral a part of his livelihood, too common a thing for him to think to mention.

Henry Thoreau seems not to be alone in forgetting about the pencil. A shop in London specializes in old carpenter's tools. There are tools everywhere, from floor to ceiling and spilling out of baskets on the sidewalk outside. The shop seems to have an example of every kind of saw used in recent centuries; there are shelves of braces and bins of chisels and piles of levels and rows of planes—everything for the carpenter, or so it seems. What the shop does not have, however, are old carpenter's pencils, items that once got equal billing in Thoreau & Company advertisements with drawing pencils for artists and engineers. The implement that was necessary to draw sketches of the carpentry job, to figure the quantities of materials needed, to mark the length of wood to be cut, to indicate the locations of holes to be drilled, to highlight the edges of wood to be planed, is nowhere to be seen. When asked where he keeps the pencils, the shopkeeper replies that he does not think there are any about. Pencils, he admits, are often found in the tool-boxes acquired by the shop, but they are thrown out with the sawdust.

In an American antique shop that deals in, among other things, old scientific and engineering instruments, there is a grand display of polished brass microscopes, telescopes, levels, balances, and scales; there are the precision instruments of physicians, navigators, surveyors, draftsmen, and engineers. The shop also has a collection of old jewelry and silverware and, behind the saltcellars, some old mechanical pencils, which appear to be there for their metal and mystery and not their utility. There are a clever Victorian combination pen and pencil in a single slender, if ornate, gold case; an unassuming little tube of brass less than two inches long that telescopes out to become a mechanical pencil of twice that length; a compact silver pencil case containing points in three colors—black, red, and blue—that can be slid into writing position; and a heavy silver pencil case that hides the half-inch stub of a still-sharpened yellow pencil of high quality. The shopkeeper will proudly show how all these work, but when asked if she has any plain wood-cased drawing pencils that the original owners of the drafting instruments must certainly have used, she will confess that she would not even know what distinguished a nineteenth-century pencil from any other kind.

Not only shops that purport to trade in the past but also museums that ostensibly preserve and display the past can seem to forget or merely ignore the indispensable role of simple objects like the pencil. Recently the Smithsonian Institution's National Museum of American History produced "After the Revolution: Everyday Life in America, 1780–1800," and one group of exhibits in the show consisted of separate worktables on which were displayed the tools of many crafts of the period: cabinet-maker and chairmaker, carpenter and joiner, shipwright, cooper, wheelwright, and others. Besides tools, many of the displays included pieces of work in progress, and a few even had wood shavings scattered about the work space, to add a sense of authenticity. Yet there was not a pencil to be seen.

© Ermakova Polina/Shutterstock.com

While many early American craftsmen would have used sharp-pointed metal scribers to mark their work, pencils would also certainly have been used when they were available. And although there was no domestic pencil industry in America in the years immediately following the Revolution, that is not to say that pencils could not be gotten. A father, writing in 1774 from England to his daughter in what were still the colonies, sent her "one dozen Middleton's best Pencils," and in the last part of the century, even after the Revolution, English pencils like Middleton's were regularly advertised for sale in the larger cities. Imported pencils or homemade pencils fashioned from reclaimed pieces of broken lead would have been the proud possessions of wood-workers especially, for carpenters, cabinetmakers, and joiners possessed the craft skill to work wood into a form that could hold pieces of graphite in a comfortable and useful way. Not only would early American woodworkers have known about, admired, wanted to possess, and tried to imitate European pencils, but also they would have prized and cared for them as they prized and cared for the kinds of tools displayed two centuries later in the Smithsonian.

These stories of absence are interesting not so much because of what they say about the lowly status of the wood-cased pencil as an artifact as because of what they say about our awareness of and our attitudes toward common things, processes, events, or even ideas that appear to have little intrinsic, permanent, or special value. An object like the pencil is generally considered unremarkable, and it is taken for granted. It is taken for granted because it is abundant, inexpensive, and as familiar as speech.

Yet the pencil need be no cliché. It can be as powerful a metaphor as the pen, as rich a symbol as the flag. Artists have long counted the pencil among the tools of their trade, and have even identified with the drawing medium. Andrew Wyeth described his pencil as a fencer's foil; Toulouse-Lautrec said of himself, "I am a pencil"; and the Moscow-born Paris illustrator and caricaturist Emmanuel Poiré took his pseudonym from the Russian word for pencil, *karandash*. In turn, the Swiss pencil-making firm of Caran d'Ache was named after this artist, and a stylized version of his signature is now used as a company logo.

The pencil, the tool of doodlers, stands for thinking and creativity, but at the same time, as the toy of children, it symbolizes spontaneity and immaturity. Yet the pencil's graphite is also the ephemeral medium of thinkers, planners, drafters, architects, and engineers, the medium to be erased, revised, smudged, obliterated, lost—or inked over. Ink, on the other hand, whether in a book or on plans or on a contract, signifies finality and supersedes the pencil drafts and sketches. If early pencilings interest collectors, it is often because of their association with the permanent success written or drawn in ink. Unlike graphite, to which paper is like sandpaper, ink flows smoothly and fills in the nooks and crannies of creation. Ink is the cosmetic that ideas will wear when they go out in public. Graphite is their dirty truth.

A glance at the index to any book of familiar quotations will corroborate the fact that there are scores of quotations extolling the pen for every one, if that, mentioning the pencil. Yet, while the conventional wisdom may be that the pen is mightier than the sword, the pencil has come to be the weapon of choice of those wishing to make better pens as well as better swords. It is often said that "everything begins with a pencil," and indeed it is the preferred medium of designers. In one recent study of the nature of the design process, engineers balked when they were asked to record their thought processes with a pen. While the directors of the study did not want the subjects to be able to erase their false starts or alter their records of creativity, the engineers did not feel comfortable or natural without a pencil in their hands when asked to comment on designing a new bridge or a better mousetrap.

Leonardo da Vinci seems to have wished to make a better everything, as his notebooks demonstrate. And when he wanted to set down his ideas for some new device, or when he merely wanted to record the state of the art of Renaissance engineering, he employed a drawing. Leonardo also used drawings to preserve his observations of natural facts, artifacts, and assorted phenomena, and he even sketched his own hand sketching. This sketch is usually identified as Leonardo's left hand, consistent with the widely held belief that the genius was left-handed. This trait in turn has been given as a reason for his mirror writing. However, it has also been convincingly argued that Leonardo was basically right-handed and was forced to use his left hand because his right was crippled in an accident. Thus Leonardo's sketch may really be of his maimed right hand as seen in a mirror by the artist drawing with his fully functioning left hand. The shortened and twisted middle finger in the sketch supports this view.

The precise nature of the drawing instrument in Leonardo's hand may also be open to some interpretation, but it appears most likely to be a small brush known from Roman times as a pencil. The lead pencil as we know it today does not seem to have existed in Leonardo's lifetime (1452–1519). Some of his sketches were done in metal point, but drawing with a pointed rod of silver or some alloy usually had to done on specially coated paper so that an otherwise faint mark would be enhanced. Some drawings were first outlined in metalpoint and then more or less traced over with a pen or a fine-pointed brush dipped in ink. This was the only kind of pencil Leonardo knew.

Name ______________________ **Course** ______________ **Date** ________________

Chapter Questions

1. "The Pencil" provides insight into the everyday items that we use but we seem to take for granted. What other items, that are oftentimes overlooked and taken for granted, are really important tools for everyday life?

2. Research the history of one of these overlooked objects and share your findings with the class.

THE 95 THESES

Martin Luther

Johann Gutenberg is credited with inventing the printing press around 1440. Before the printing press, written materials were rare and costly because they had to be drafted and copied by hand. The only people to have books were the wealthy and powerful, so, traditionally, the lower classes accepted the judgment and rules of authority figures because information was not able to circulate to anyone else. One of the first ways that the printing press tremendously impacted society occurred in 1517-18 when Martin Luther's 95 Theses was mass printed. Many argue that this list of criticisms of the Roman Catholic Church was a major cause of the Protestant Reformation (a major 16th century European movement aimed at reexamining and changing the beliefs and policies of the Church).

The Ninety-Five Theses, by Martin Luther, first published 1517 (print), German School, (16th century) / Deutsches Historisches Museum, Berlin, Germany / Bridgeman Images.

OCTOBER 31, 1517

Out of love for the truth and the desire to bring it to light, the following propositions will be discussed at Wittenberg, under the presidency of the Reverend Father Martin Luther, Master of Arts and of Sacred Theology, and Lecturer in Ordinary on the same at that place.

Wherefore he requests that those who are unable to be present and debate orally with us, may do so by letter.

In the Name our Lord Jesus Christ. Amen.

1. Our Lord and Master Jesus Christ, when He said Poenitentiam agite, willed that the whole life of believers should be repentance.

2. This word cannot be understood to mean sacramental penance, i.e., confession and satisfaction, which is administered by the priests.

3. Yet, it means not inward repentance only; nay, there is no inward repentance which does not outwardly work divers mortifications of the flesh.

4. The penalty [of sin], therefore, continues so long as hatred of self continues; for this is the true inward repentance, and continues until our entrance into the kingdom of heaven.

5. The Pope does not intend to remit, and cannot remit any penalties other than those which he has imposed either by his own authority or by that of the Canons.

6. The Pope cannot remit any guilt, except by declaring that it has been remitted by God and by assenting to God's remission; though, to be sure, he may grant remission in cases reserved to his judgment. If his right to grant remission in such cases were despised, the guilt would remain entirely unforgiven.

7. God remits guilt to no one whom He does not, at the same time, humble in all things and bring into subjection to His vicar, the priest.

8. The penitential canons are imposed only on the living, and, according to them, nothing should be imposed on the dying.

9. Therefore the Holy Spirit in the Pope is kind to us, because in his decrees he always makes exception of the article of death and of necessity.

10. Ignorant and wicked are the doings of those priests who, in the case of the dying, reserve canonical penances for purgatory.

This picture from 1881 shows Johann Gutenberg (right), the inventor of the printing press, working on an engraving.

11. This changing of the canonical penalty to the penalty of purgatory is quite evidently one of the tares that were sown while the bishops slept.

12. In former times the canonical penalties were imposed not after, but before absolution, as tests of true contrition.

13. The dying are freed by death from all penalties; they are already dead to canonical rules, and have a right to be released from them.

14. The imperfect health [of soul], that is to say, the imperfect love, of the dying brings with it, of necessity, great fear; and the smaller the love, the greater is the fear.

15. This fear and horror is sufficient of itself alone (to say nothing of other things) to constitute the penalty of purgatory, since it is very near to the horror of despair.

16. Hell, purgatory, and heaven seem to differ as do despair, almost-despair, and the assurance of safety.

17. With souls in purgatory it seems necessary that horror should grow less and love increase.

18. It seems unproved, either by reason or Scripture, that they are outside the state of merit, that is to say, of increasing love.

19. Again, it seems unproved that they, or at least that all of them, are certain or assured of their own blessedness, though we may be quite certain of it.

20. Therefore by "full remission of all penalties" the Pope means not actually "of all," but only of those imposed by himself.

21. Therefore those preachers of indulgences are in error who say that by the Pope's indulgences a man is freed from every penalty, and saved.

22. Whereas he remits to souls in purgatory no penalty which, according to the canons, they would have had to pay in this life.

23. If it is at all possible to grant to any one the remission of all penalties whatsoever, it is certain that this remission can be granted only to the most perfect, that is, to the very fewest.

24. It must needs be, therefore, that the greater part of the people are deceived by that indiscriminate and high-sounding promise of release from penalty.

25. The power which the Pope has, in a general way, over purgatory, is just like the power which any bishop or curate has, in a special way, within his own diocese or parish.

26. The Pope does well when he grants remission to souls (in purgatory), not by the power of the keys (which he does not possess), but by way of intercession.

27. They preach man who say that so soon as the penny jingles into the money-box, the soul flies out (of purgatory).

28. It is certain that when the penny jingles into the money-box, gain and avarice can be increased, but the result of the intercession of the Church is in the power of God alone.

29. Who knows whether all the souls in purgatory wish to be bought out of it, as in the legend of Sts. Severinus and Paschal.

30. No one is sure that his own contrition is sincere; much less that he has attained full remission.

31. Rare as is the man that is truly penitent, so rare is also the man who truly buys indulgences, i.e., such men are most rare.

32. They will be condemned eternally, together with their teachers, who believe themselves sure of their salvation because they have letters of pardon.

33. Men must be on their guard against those who say that the Pope's pardons are that inestimable gift of God by which man is reconciled to Him.

34. For these "graces of pardon" concern only the penalties of sacramental satisfaction, and these are appointed by man.

35. They preach no Christian doctrine who teach that contrition is not necessary in those who intend to buy souls out of purgatory or to buy confessionalia.

36. Every truly repentant Christian has a right to full remission of penalty and guilt, even without letters of pardon.

37. Every true Christian, whether living or dead, has part in all the blessings of Christ and the Church; and this is granted him by God, even without letters of pardon.

38. Nevertheless, the remission and participation (in the blessings of the Church) which are granted by the Pope are in no way to be despised, for they are, as I have said, the declaration of divine remission.

39. It is most difficult, even for the very keenest theologians, at one and the same time to commend to the people the abundance of pardons and (the need of) true contrition.

40. True contrition seeks and loves penalties, but liberal pardons only relax penalties and cause them to be hated, or at least, furnish an occasion (for hating them).

41. Apostolic pardons are to be preached with caution, lest the people may falsely think them preferable to other good works of love.

42. Christians are to be taught that the Pope does not intend the buying of pardons to be compared in any way to works of mercy.

43. Christians are to be taught that he who gives to the poor or lends to the needy does a better work than buying pardons.

44. Because love grows by works of love, and man becomes better; but by pardons man does not grow better, only more free from penalty.

45. Christians are to be taught that he who sees a man in need, and passes him by, and gives (his money) for pardons, purchases not the indulgences of the Pope, but the indignation of God.

46. Christians are to be taught that unless they have more than they need, they are bound to keep back what is necessary for their own families, and by no means to squander it on pardons.

47. Christians are to be taught that the buying of pardons is a matter of free will, and not of commandment.
48. Christians are to be taught that the Pope, in granting pardons, needs, and therefore desires, their devout prayer for him more than the money they bring.
49. Christians are to be taught that the Pope's pardons are useful, if they do not put their trust in them; but altogether harmful, if through them they lose their fear of God.
50. Christians are to be taught that if the Pope knew the exactions of the pardon-preachers, he would rather that St. Peter's church should go to ashes, than that it should be built up with the skin, flesh, and bones of his sheep.
51. Christians are to be taught that it would be the Pope's wish, as it is his duty, to give of his own money to very many of those from whom certain hawkers of pardons cajole money, even though the church of St. Peter might have to be sold.
52. The assurance of salvation by letters of pardon is vain, even though the commissary, nay, even though the Pope himself, were to stake his soul upon it.
53. They are enemies of Christ and of the Pope, who bid the Word of God be altogether silent in some Churches, in order that pardons may be preached in others.
54. Injury is done the Word of God when, in the same sermon, an equal or a longer time is spent on pardons than on this Word.
55. It must be the intention of the Pope that if pardons, which are a very small thing, are celebrated with one bell, with single processions and ceremonies, then the Gospel, which is the very greatest thing, should be preached with a hundred bells, a hundred processions, a hundred ceremonies.
56. The "treasures of the Church," out of which the Pope grants indulgences, are not sufficiently named or known among the people of Christ.
57. That they are not temporal treasures is certainly evident, for many of the vendors do not pour out such treasures so easily, but only gather them.
58. Nor are they the merits of Christ and the Saints, for even without the Pope, these always work grace for the inner man, and the cross, death, and hell for the outward man.
59. St. Lawrence said that the treasures of the Church were the Church's poor, but he spoke according to the usage of the word in his own time.
60. Without rashness we say that the keys of the Church, given by Christ's merit, are that treasure.
61. For it is clear that for the remission of penalties and of reserved cases, the power of the Pope is of itself sufficient.
62. The true treasure of the Church is the Most Holy Gospel of the glory and the grace of God.
63. But this treasure is naturally most odious, for it makes the first to be last.

64. On the other hand, the treasure of indulgences is naturally most acceptable, for it makes the last to be first.

65. Therefore the treasures of the Gospel are nets with which they formerly were wont to fish for men of riches.

66. The treasures of the indulgences are nets with which they now fish for the riches of men.

67. The indulgences which the preachers cry as the "greatest graces" are known to be truly such, in so far as they promote gain.

68. Yet they are in truth the very smallest graces compared with the grace of God and the piety of the Cross.

69. Bishops and curates are bound to admit the commissaries of apostolic pardons, with all reverence.

70. But still more are they bound to strain all their eyes and attend with all their ears, lest these men preach their own dreams instead of the commission of the Pope.

71. He who speaks against the truth of apostolic pardons, let him be anathema and accursed!

72. But he who guards against the lust and license of the pardon-preachers, let him be blessed!

73. The Pope justly thunders against those who, by any art, contrive the injury of the traffic in pardons.

74. But much more does he intend to thunder against those who use the pretext of pardons to contrive the injury of holy love and truth.

75. To think the papal pardons so great that they could absolve a man even if he had committed an impossible sin and violated the Mother of God—this is madness.

76. We say, on the contrary, that the papal pardons are not able to remove the very least of venial sins, so far as its guilt is concerned.

77. It is said that even St. Peter, if he were now Pope, could not bestow greater graces; this is blasphemy against St. Peter and against the Pope.

78. We say, on the contrary, that even the present Pope, and any Pope at all, has greater graces at his disposal; to wit, the Gospel, powers, gifts of healing, etc., as it is written in I. Corinthians xii.

79. To say that the cross, emblazoned with the papal arms, which is set up [by the preachers of indulgences], is of equal worth with the Cross of Christ, is blasphemy.

80. The bishops, curates, and theologians who allow such talk to be spread among the people, will have an account to render.

81. This unbridled preaching of pardons makes it no easy matter, even for learned men, to rescue the reverence due to the Pope from slander, or even from the shrewd questionings of the laity.

82. To wit:—"Why does not the Pope empty purgatory, for the sake of holy love and of the dire need of the souls that are there, if he redeems an infinite number of souls for the sake of miserable money with which to build a Church? The former reasons would be most just; the latter is most trivial."

83. Again:—"Why are mortuary and anniversary masses for the dead continued, and why does he not return or permit the withdrawal of the endowments founded on their behalf, since it is wrong to pray for the redeemed?"

84. Again:—"What is this new piety of God and the Pope, that for money they allow a man who is impious and their enemy to buy out of purgatory the pious soul of a friend of God, and do not rather, because of that pious and beloved soul's own need, free it for pure love's sake?"

85. Again:—"Why are the penitential canons long since in actual fact and through disuse abrogated and dead, now satisfied by the granting of indulgences, as though they were still alive and in force?"

86. Again:—"Why does not the Pope, whose wealth is today greater than the riches of the richest, build just this one church of St. Peter with his own money, rather than with the money of poor believers?"

87. Again:—"What is it that the Pope remits, and what participation does he grant to those who, by perfect contrition, have a right to full remission and participation?"

88. Again:—"What greater blessing could come to the Church than if the Pope were to do a hundred times a day what he now does once, and bestow on every believer these remissions and participations?"

89. "Since the Pope, by his pardons, seeks the salvation of souls rather than money, why does he suspend the indulgences and pardons granted heretofore, since these have equal efficacy?"

90. To repress these arguments and scruples of the laity by force alone, and not to resolve them by giving reasons, is to expose the Church and the Pope to the ridicule of their enemies, and to make Christians unhappy.

91. If, therefore, pardons were preached according to the spirit and mind of the Pope, all these doubts would be readily resolved; nay, they would not exist.

92. Away, then, with all those prophets who say to the people of Christ, "Peace, peace," and there is no peace!

93. Blessed be all those prophets who say to the people of Christ, "Cross, cross," and there is no cross!

94. Christians are to be exhorted that they be diligent in following Christ, their Head, through penalties, deaths, and hell.

95. And thus be confident of entering into heaven rather through many tribulations, than through the assurance of peace.

Name ______________________ **Course** ____________ **Date** _______________

Chapter Questions

1. What impact did Gutenberg's printing press have on the Protestant Reformation?

2. How did written information travel prior to this invention of the printing press?

3. What effect did print have on society in the 1500s, when Martin Luther wrote The 95 Theses, and what effect does printing have on people today?

THE TELEPHONE

Anwar F. Accawi

People do not often acknowledge the impact of modern "everyday" technologies, like landline telephones, on society. Author, Anewar Accawi, was raised in the rural Lebanese village of Madaluna. In 1965, Accawi came to the United States on a college scholarship and has written short stories and books regarding his life in this village. One of his pieces, "The Telephone," vividly describes the extreme impact the first telephone had on this community and how people changed as a result. Before modern technologies, people judged time by certain events. Children would happily run errands for adults for small sums of money. After the phone arrives, however, Accawi discusses how life changes physically, emotionally, and financially for the members of the community.

© Brendan Howard/Shutterstock.com

When I was growing up in Magdaluna, a small Lebanese village in the terraced, rocky mountains east of Sidon, time didn't mean much to anybody, except maybe to those who were dying, or those waiting to appear in court because they had tampered with the boundary markers on their land. In those days, there was no real need for a calendar or a watch to keep track of the hours, days, months, and years. We knew what to do and when to do it, just as the Iraqi geese knew when to fly north, driven by the hot wind that blew in from the desert, and the ewes knew when to give birth to wet lambs that stood on long, shaky legs in the chilly March wind and baaed hesitantly, because they were small and cold and did not know where they were or what to do now that they were here. The only timepiece we had need of then was the sun. It rose and set, and the seasons rolled by, and we sowed seed and harvested and ate and played and married our cousins and had babies who got whooping cough and chickenpox—and those children who survived grew up and married *their* cousins and had babies who got whooping cough and chickenpox. We lived and loved and toiled and died without ever needing to know what year it was, or even the time of day.

It wasn't that we had no system for keeping track of time and of the important events in our lives. But ours was a natural—or rather, a divine—calender, because it was framed by acts of God Allah himself set down the milestones with earthquakes and droughts and floods and locusts and pestilences. Simple as our calendar was, it worked just fine for us.

Take, for example, the birth date of Teta Im Khalil, the oldest woman in Magdaluna and all the surrounding villages. When I first met her, we had just returned home from Syria at the end of the Big War and were living with Grandma Mariam. Im Khalil came by to welcome my father home and to take a long, myopic look at his foreign-born wife, my mother. Im Khalil was so old that the skin of her cheeks looked like my father's grimy tobacco pouch, and when I kissed her (because Grandma insisted that I show her old friend affection), it was like kissing a soft suede glove that had been soaked with sweat and then left in a dark closet for a season. Im Khalil's face got me to wondering how old one had to be to look and taste the way she did. So, as soon as she had hobbled off on her cane, I asked Grandma, "How old is Teta Im Khalil?"

Grandma had to think for a moment; then she said, "I've been told that Teta was born shortly after the big snow that caused the roof on the mayor's house to cave in."

"And when was that?" I asked.

"Oh, about the time we had the big earthquake that cracked the wall in the east room."

Well, that was enough for me. You couldn't be more accurate than that, now, could you? Satisfied with her answer, I went back to playing with a ball made from an old sock stuffed with other, much other socks.

And that's the way it was in our little village for as far back as anybody could remember: people were born so many years before or after an earthquake or a flood; they got married or died so many years before or after a long drought or a big snow or some other disaster. One of the most unusual of these dates was when Antoinette the seamstress and Saeed the barber (and tooth puller) got married. That was the year of the whirlwind during which fish and oranges fell from the sky. Incredible as it may sound, the story of the fish and oranges was true, because men—respectable men, like Abu George the blacksmith and Abu Asaad the mule skinner, men who would not lie even to save their own souls—told and retold that story until it was incorporated into Magdaluna's calendar, just like the year of the black moon and the year of the locusts before it. My father, too, confirmed the story for me. He told me that he had been a small boy himself when it had rained fish and oranges from heaven. He'd gotten up one morning after a stormy night and walked out into the yard to find fish as long as his forearm still flopping here and there among the wet navel oranges.

The year of the fish-bearing twister, however, was not the last remarkable year. Many others followed in which strange and wonderful things happened: milestones added by the hand of Allah to Magdaluna's calendar. There was, for instance, the year of the drought, when the heavens were shut for months and the spring from which the entire village got its drinking water slowed to a trickle. The spring was about a mile from the village, in a ravine that opened at one end into a small, flat clearing covered with fine gray dust and hard, marble-sized goat droppings, because every afternoon the goatherds brought their flocks there to water them. In the year of the drought, that little clearing was always packed full of noisy kids with big brown eyes and sticky hands, and their months—sinewy, overworked young women with protruding collarbones and cracked, callused brown heels. The children ran around playing tag or hide-and-seek while the while women talked, shooed flies, and awaited their turns to fill up their jars with drinking water to bring home to their napping men and wet habies. There were days when we had to wait from sunup until late afternoon just to fill a small clay jar with precious, cool water.

Sometimes, amid the long wait and the heat and the flies and the smell of goat dung, tempers flared, and the younger women, anxious about their babies, argued over whose turn it was to fill up her jar. And sometimes the arguments escalated into full-blown, khockdown-dragout fights; the women would grab each other by the hair and curse and scream and spit and call each other names that made my ears tingle. We little brown boys who went with our mothers to fetch water loved these fights, because we got to see the women's legs and their colored panties as they grappled and rolled around in the dust. Once in a while, we got lucky and saw much more, because some of the women wore nothing at all under their long dresses. God, how I used to look forward to those fights. I remember the rush, the excitement, the sun dancing on the dust clouds as a dress ripped and a young white breast was revealed, then quickly hidden. In my calendar, that year of drought will always be one of the best years of my childhood, because it was then, in a dusty clearing by a trickling mountain spring, I got my first glimpses of the wonders, the mysteries, and the promises hidden beneath the folds of a woman's dress. Fish and oranges from heaven . . . you can get over that.

But, in another way, the year of the drought was also one of the worst of my life, because that was the year that Abu Raja, the retired cook who used to entertain us kids by cracking walnuts on his forehead, decided it was time Magdaluna got its own telephone. Every civilized village needed a telephone, he said, and Magdaluna was not going to get anywhere until it had one. A telephone would link us with the outside world. At the time, I was too young to understand the debate, but a few men—like Shukri, the retired Turkish-army drill sergeant, and Abu Hanna the vineyard keeper—did all they could to talk Abu Raja out of having a telephone brought to the village. But they were outshouted and ignored and finally shunned by the other villagers for resisting progress and trying to keep a good thing from coming to Magdaluna.

One warm day in early fall, many of the villagers were out in their fields repairing walls or gathering wood for the winter when the shout went out that the telephone-company truck had arrived at Abu Raja's *dikkan*, or country store. There were no roads in those days, only footpaths and dry steambeds, so it took the telephone-company truck almost a day to work its way up the rocky terrain from Sidon—about the same time it took to walk. When the truck came into view, Abu George, who had a huge voice and, before the telephone, was Magdaluna's only long-distance communication system, bellowed the news from his front porch. Everybody dropped what they were doing and ran to Abu Raja's house to see what was happening. Some of the more dignified villagers, however, like Abu Habeeb and Abu Nazim, who had been to big cities like Beirut and Damascus and had seen things like telephones and telegraphs, did not run the way the rest did; they walked with their canes hanging from the crooks of their arms, as if on a Sunday afternoon stroll.

It did not take long for the whole village to assemble at Abu Raja's *dikkan*. Some of the rich villagers, like the widow Farha and the gendarme Abu Nadeem, walked right into the store and stood at the elbows of the two important-looking men from the telephone company, who proceeded with utmost gravity, like priests at Communion, to wire up the telephone. The poorer villagers stood outside and listened carefully to the details relayed to them by the not-so-poor people who stood in the doorway and could see inside.

"The bald man is cutting the blue wire," someone said.

"He is sticking the wire into the hole in the bottom of the black box," someone else added.

"The telephone man with the mustache is connecting two pieces of wire. Now he is twisting the ends together," a third voice chimed in.

Because I was small and unaware that I should have stood outside with the other poor folk to give the rich people inside more room (they seemed to need more of it than poor people did), I wriggled my way through the dense forest of legs to get a first-hand look at the action. I felt like the barefoot Moses, sandals in hand, staring at the burning bush on Mount Sinai. Breathless, I watched as the men in blue, their shirt pockets adorned with fancy lettering in a foreign language, put together a black machine that supposedly would make it possible to talk with uncles, aunts, and cousins who lived more than two days' ride away.

It was shortly after sunset when the man with the mustache announced that the telephone was ready to use. He explained that all Abu Raja had to do was lift the receiver, turn the crank on the black box a few times, and wait for an operator to take his call. Abu Raja, who had once lived and worked in Sidon, was impatient with the telephone man for assuming that he was ignorant. He grabbed the receiver and turned the crank forcefully, as if trying to start a Model T Ford. Everybody was impressed that he knew what to do. He even called the operator by her first name. "Centralist." Within moments, Abu Raja was talking with his brother, a concierge in Beirut. He didn't even have to raise his voice or shout to be heard.

If I hadn't seen it with my own two eyes and heard it with my own two ears, I would not have believed it—and my friend Kameel didn't. He was away that day watching his father's goats, and when he came back to the village that evening, his cousin Habeen and I told him about the telephone and how Abu Raja had used it to speak with his brother in Beirut. After he heard our report, Kameel made the sign of the cross, kissed his thumbnail, and warned us that lying was a bad sin and would surely land us in purgatory. Kameel believed in Jesus and Mary, and wanted to be a priest when he grew up. He always crossed himself when Habeeb, who was irreverent, and I, who was Presbyterian, were around, even when we were not bearing bad news.

And the telephone, as it turned out, was bad news. With its coming, the face of the village began to change. One of the first effects was the shifting of the village's center. Before the telephone's arrival, the men of the village used to gather regularly at the house of Im Kaleem, a short, middle-aged widow with jet-black hair and a raspy voice that could be heard all over the village, even when she was only whispering. She was a devout Catholic and also the village *shlikki*—whore. The men met at her house to argue about politics and drink coffee and play cards or backgammon. Im Kaleem was not a true prostitute, however, because she did not charge for her services—not even for the coffee and tea (and, occasionally, the strong liquor called arrack) that she served the men. She did not need the money; her son, who was overseas in Africa, sent her money regularly. (I knew this because my father used to read her son's letters to her and take down her replies, as Im Kaleem could not read and write.) Im Kaleem was no slut either—unlike some women in the village—because she loved all the men she entertained, and they loved her, every one of them. In a way, she was married to all the men in the village. Everybody knew it—the wives knew it; the itinerant Catholic priest knew it; the Presbyterian minister knew it—but nobody objected. Actually, I suspect the women (my mother included) did not mind their husband's visits to Im Kaleem. Oh, they wrung their hands and complained to one another about their men's unfaithfulness, but secretly they were relieved, because Im Kaleem took some of the pressure off them and kept the men out of their hair while they attended to their endless chores. Im Kaleem was also a kind of confessor and troubleshooter, talking sense to those men who were having family problems, especially the younger ones.

Before the telephone came to Magdaluna, Im Kaleem's house was bustling at just about any time of day, especially at night, when its windows were brightly lit with three large oil lamps, and the loud voices of the men talking, laughing, and arguing could be heard in the street below—a reassuring, homey sound. Her house was an island of comfort, an oasis for the weary village men, exhausted from having so little to do.

But it wasn't long before many of those men—the younger ones especially—started spending more of their days and evenings at Abu Raja's *dikkan*. There, they would eat and drink and talk and play checkers and backgammon, and then lean their chairs back against the wall—the signal that they were ready to toss back and forth, like a ball, the latest rumors going around the village. And they were always looking up from their games and drinks and talk to glance at the phone in the corner, as if expecting it to ring any minute and bring news that would change their lives and deliver them from their aimless existence. In the meantime, they smoked cheap, hand-rolled cigarettes, dug dirt out from under their fingernails with big pocketknives, and drank lukewarm sodas that they called Kacula, Seffen-Ub, and Bebsi. Sometimes, especially when it was hot, the days dragged on so slowly that the men turned on Abu Saeed, a confirmed bachelor who practically lived in Abu Raja's *dikkan*, and teased him for going around barefoot and unshaven since the Virgin has appeared to him behind the olive press.

The telephone was also bad news for me personally. It took away my lucrative business—a source of much-needed income. Before the telephone came to Magdaluna, I used to hang around Im Kaleem's courtyard and play marbles with the other kids, waiting for some man to call down from a window and ask me to run to the store for cigarettes or arrack, or to deliver a message to his wife, such as what he wanted for supper. There was always something in it for me: a ten-or even a twenty-five-piaster piece. On a good day, I ran nine or ten of those errands, which assured a steady supply of marbles that I usually lost to Sami or his cousin Hani, the basket weaver's boy. But as the days went by, fewer and fewer men came to Im Kaleem's, and more and more congregated at Abu Raja's to wait by the telephone. In the evenings, no light fell from her window onto the street below, and the laughter and noise of the men trailed off and finally stopped. Only Shukri, the retired Turkish-army drill sergeant, remained faithful to Im Kaleem after all the other men had deserted her; he was still seen going into or leaving her house from time to time. Early that winter, Im Kaleem's hair suddenly turned turned gray, and she got sick and old. Her legs started giving her trouble, making it hard for her to walk. By spring she hardly left her house anymore.

At Abu Raja's *dikkan*, the calls did eventually come, as expected, and men and women started leaving the village the way a hailstorm begins: first one, then two, then bunches. The army took them. Jobs in the cities lured them. And ships and airplanes carried them to such faraway places as Australia and Brazil and New Zealand. My friend Kameel, his cousin Habeeb, and their cousins and my cousins all went away to become ditch diggers and mechanics and butcher-shop boys and deli owners who wore dirty aprons sixteen hours a day; all looking for a better life than the one they had left behind. Within a year, only the sick, the old, and the maimed were left in the village. Magdaluna became a skeleton of its former self, desolate and forsaken, like the tombs, a place to get away from.

Finally, the telephone took my family away, too. My father got a call from an old army buddy who told him that an oil company in southern Lebanon was hiring interpreters and instructors. My father applied for a job and got it, and we moved to Sidon, where I went to a Presbyterian missionary school and graduated in 1962. Three years later, having won a scholarship, I left Lebanon for the United States. Like the others who left Magdaluna before me, I am still looking for that better life.

Name ______________________ Course ____________ Date _______________

Chapter Questions

1. "The Telephone" demonstrates how people rely on technology to do a task, even though they once carried on sufficiently without it. What other technological advancements, besides telephones, have drastically changed the ways that certain tasks are carried out in modern society? In these examples, do you think that people have become too reliant on these technologies to do simple tasks?

2. From personal experience, what positive effects (if any) come from being away from modern technologies such as telephones or cell phones?

RADIO

Brent D. Ruben, Paul Reis, Barbara Iverson, and Genelle I. Belmas

© DmitriMaruta/Shutterstock.com

Introduction

Radio is one of the oldest electronic mass medium in society. Over time, technology and public policy has caused radio's function and purpose to change. What was once just a few bleeps sent from one ship to another is now crystal-clear voices communicating around the world. Although radio waves float freely though the air, the contents of those waves are highly restricted. There are government regulations about who can own radio stations and who controls subject matter over the airwaves, in addition to economic models that dictate who pays and who makes money. These factors help determine how radio fits into a society.

In commercial radio, what news, information, and entertainment the audience wishes to hear and how they choose to receive that content have an influential role in how radio functions as well. The attention of an audience did—and can—create a "golden age" for the medium. However, just as quickly, a shift in attention can end one.

To talk about radio, we need to know that **radio**, from a technological standpoint, means telecommunication modulated by electromagnetic waves. However, *radio* also may refer to radio programming, modern music radio, the radio industry, or radio stations in general. All these meanings of *radio* inform our understanding of radio as a mass medium.

Today there are more than 16,000 AM radio stations, 26,000 FM stations, and 1500 shortwave radio stations in the world. In addition, over 3 million people subscribe to satellite radio. As we'll explore in this chapter, modern radio only vaguely resembles the original concept of wireless electromagnetic transmissions. These days, people are as likely to be listening to what they call "radio" on an MP3 player, a computer, or via satellite transmission in their cars as they are to be listening to an AM/FM receiver.

A Historical Overview

It's hard to believe that what started as a few taps of Morse code in 1901 grew into the diverse multimedia outlet that is today's radio industry. Indeed, the history of radio is an ongoing story of vision, ingenuity, and rapid technological advancement.

Birth of a Medium—1888–1912

Although radio as we know it today has only been around for a little over a century, radio waves have been present as long as planets have orbited the sun. However, it took the innovative thinking of many scientists and inventors to make sense of these invisible waves. German physicist Heinrich Hertz was the first to prove the existence of radio waves in the late 1880s. His experiments proved that, like light, radio waves are a type of electromagnetic wave.[1] **Electromagnetic waves** are movements of energy that can travel at the speed of light in a free space detached from wires. Soon after Hertz made these discoveries, scientists were eager to find a way to use radio waves to carry sounds, including voices.

As a young man, Guglielmo Marconi read about Hertz's work and began conducting his own experiments in a quest to develop wireless telegraphy. By 1901, Marconi had transmitted the first radio message across the Atlantic Ocean—Morse code for the letter "S." Over the next several years, radio came to be used as a form of ship-to-shore communication. In 1912, wireless operators were onboard both the *Titanic* and the ship that rescued its survivors, the *Carpathia*. Marconi's technology allowed radio operators to transmit information about the disaster and ask for assistance before the *Carpathia* reached its port in New York City.[2]

Meanwhile, Lee de Forest was developing his own innovations for radio, including the Audion, a device that he patented in 1907. De Forest's device was used to convert radio frequency into audio frequency, so sounds could be transmitted and amplified. The Audion made live broadcasting possible and later became a key component of all telephone, television, and radar systems before the invention of the transistor.[3] Although some admirers refer to de Forest as the "father of radio," some critics of his time claimed that he merely patented many ideas that others had generated.[4]

---

[1] "Heinrich Rudolf Hearst," Physicists' biographies at http://phisicist.info/hertz.html. Accessed July 17, 2008.

[2] Helen Briggs, "Profile: Marconi, the wireless pioneer," *BBC News Online*, December 11, 2001, at http://news.bbc.co.uk/2/hi/science/nature/1702037.stm.

[3] "de Forest, Lee," *Encyclopedia Britannica*, at http://search.eb.com/eb/article-9029588. Accessed July 21, 2008.

[4] "The Complete Lee de Forest," *Perham Collection*, History San Jose, 2003, at http://www.leedeforest.org. Accessed July 17, 2008.

Ironically, de Forest brought similar claims against another pioneer of early radio, Edwin Armstrong. In 1912, Armstrong took de Forest's invention of the Audion one step further by devising a way to improve the amplification by a thousandfold, so the sound generated could be heard across a room. The most impressive part of Armstrong's take on the Audion—later named the regenerative circuit—was that it could serve as an oscillator as well as a receiver. This allowed the device to both generate and obtain radio waves. Shortly after receiving a patent for his idea, he crossed paths with de Forest, who claimed ownership of the idea. Although de Forest eventually won a lengthy legal battle on the basis of a language technicality, Armstrong is still credited with the invention.[5]

As illustrated by de Forest and Armstrong, the development of radio in its earliest forms depended on the ingenuity and, at times, fierce competitiveness of pioneering scientists.

From Work to Play, Tool to Toy

Radio communications were a vital link in the military chain of command during World War I. This caused the U.S. government to place heavy restrictions on radio broadcasting. At the end of World War I in 1918, many of those restrictions were lifted and radio broadcasting was ready to take on a new purpose. David Sarnoff, a Russian immigrant who had worked for the Marconi Company during the war, had a vision of radio as a home entertainment device. Sarnoff believed that radio could bring music into the home in a way no one else had imagined. He also envisioned a radio receiver in every home. Sarnoff's tenacious efforts helped him rise to fame as one of the first on-air radio personalities and a leader at Radio Corporation of America (RCA).[6]

A radio in every home was a visionary idea for a time when there was no mass communication beyond print brought directly to the individual. KDKA, the first commercial radio station, which began broadcasting in 1920, changed all that. It featured music and entertainment, but it also carried the election returns for the 1920 presidential election.

The main downfall of radio was the high operating costs. Running a broadcasting station required strong financial backing but realized little or no profit. However, as radio stations began to rise in popularity and value, businesses started to view radio as a good platform for advertising. Eventually, advertising became the primary means of support for radio broadcasting in the United States.

As its popularity grew, large corporations wanted to get into the business of radio. In 1926, Sarnoff brought together the resources of RCA, General Electric, and Westinghouse to purchase WEAF in New York. WEAF eventually became the anchor station for the National Broadcasting System (NBC). NBC soon swelled to 25 stations nationwide and became the first broadcasting network. In 1927, NBC was able to broadcast the Rose Bowl from California to New York through the use

---

[5] Armstrong, Edwin H. *Encyclopedia Britannica* at http://search.eb.com/eb/article-410. Accessed July 22, 2008.

[6] broadcasting. *Encyclopedia Britannica* at http://www.britannica.com/EBchecked/topic/80543/broadcasting. Accessed July 21, 2008.

of long-distance wire telephone lines. Fans from coast to coast could hear the football game for the first time.

Still, music was the "killer app" that Sarnoff believed would make radio a household necessity. From the beginning, the success of radio and the success of music recording industries have been intertwined. Americans bought radios so live music could be delivered directly into their homes. In turn, radio spurred the sales of records, as listeners sought out their favorite songs or bands after hearing them on the radio. Sarnoff's hunch that broadcasting and music recording would influence one another over time turned out to be accurate.[7]

Reporting and News Radio

Although much of radio programming in the 1920s and 30s revolved around music and entertainment, there was an interest in radio news as well. In addition to the 1920 KDKA broadcast of the election, news events such as presidential inaugurations, baseball's World Series, and aviator Charles Lindbergh's safe landing in Paris after his solo flight across the Atlantic were broadcast on the radio. However, regular newscasts didn't catch on until the mid-1930s. These first newscasts consisted of summaries of the headlines from the morning newspapers read over the air.

Newspaper publishers feared that radio would lure their audiences away, and many refused to print radio schedules in the early years. The newspaper publishers blocked the press associations (AP, UP, and INS) from selling news for radio broadcasts until the 1940s.

The radio networks understood the potential of broadcast news and began to create syndicates and the capability for reporting national and global events. With the start of World War II, radio became the most important link between the public, Washington, D.C., and the rest of the world. With their loved ones thousands of miles away, listeners tuned in to the familiar voices of correspondents who reported from abroad to "bring home" news of the war to the United States.

Edward R. Murrow of the Columbia Broadcasting System (CBS) was in Europe when the war began. He quickly assembled a team of reporters to report nightly to London on the progress of the war. Honest, urgent, and live at the scene, these broadcasts came to define broadcast news for decades.

In the new medium of radio, there were no models for reporters to follow. Murrow didn't have formal training as a journalist. He thought that made it easier to report for radio because he didn't have to unlearn anything before jumping into radio reporting.

One of his innovative broadcasts features the sound of British civilians entering an air-raid shelter. Murrow set the microphone on the ground to record the quiet, orderly footsteps that reinforced his words about the courage of the average citizen during the London Blitz.

---

[7] "About RCA: Linking the Nation," *RCA*, at http://home.rca.com/en-US/ PressReleaseDetail.html?Cat=RCA History&MN=7. Accessed July 16, 2008.

Murrow and his team of reporters, called his boys, set an example at CBS that became the model for all the networks. Investigative reporters and political correspondents adopted the blunt style of the reporter, microphone in hand, who steps up to politicians, generals, corporate leaders, or John Q. Public and asks the tough questions, demanding to know the truth. This style came to define hard-hitting journalism and also migrated to television with Murrow and his protégés.

From Novelty to Network

Due to the influence of World War I (1914–1918) and the Great Depression (1929–1939), the structure of radio broadcasting developed under different economic and social conditions than those that exist today. The radio industry was based on a different set of assumptions about programs, producers, technology, and consumers.

Newspapers were the dominant medium at the time of radio's introduction into the world of mass communications. Radio stations had to find **economies of scale** through the wire services that allowed companies to pool their resources in order to extend their coverage. Creating these networks also allowed stations to share the expense of producing and transmitting content. Juggernauts of the industry quickly emerged.

After NBC was created, the company formed two semi-independent radio networks, the Blue Network (WJZ) and the Red Network (WEAF). The Red Network flourished and, by 1938, carried 75 percent of NBC's commercial programs. The askew performance of the two networks prompted NBC to sell the Blue Network, which eventually became the American Broadcasting Company (ABC), in 1943.[8]

Shortly after NBC's conception, a cigar-business owner named William S. Paley combined the United Independent Broadcasters and the Columbia Phonograph Company to create the Columbia Broadcasting System (CBS). A businessman at heart, Paley strove to turn a struggling radio network into a success. The key to this accomplishment was advertising. He offered free programming to affiliated stations if those stations would designate slots of airtime to network shows that were sponsored by advertisers. Paley's gimmick helped CBS expand from 22 stations in 1928 to 144 stations over the course of a decade. Although CBS produced many successful radio shows that featured stars such as Fred Allen, Bing Crosby, and Kate Smith, the network's repertoire grew considerably after Paley's raid of NBC's talent pool. Paley was able to coerce stars such as Jack Benny, George Burns, and comedy duo Amos 'n' Andy to join the CBS network. This attack hurt NBC even more when the age of television approached, as these personalities proved to be as popular on the screen as they were on the air.[9]

---

[8] National Broadcasting Co., Inc. *Encyclopedia Britannica* at http://www.britannica.com/EBchecked/topic/404533/National-Broadcasting-Co-Inc. Accessed July 22, 2008.

[9] CBS Corporation. *Encyclopedia Britannica* at http://www.britannica.com/EBchecked/topic/100876/CBS-Corporation. Accessed July 22, 2008.

Not all radio networks were created to compete on a national level. In 1934, the Mutual Broadcasting System (MBS) was developed as a cooperative network shared by WOR in New York, WGN in Chicago, WLW in Cincinnati, and WXYZ in Detroit. Unlike the larger networks, MBS had no production studio or centralized corporate owners. The network catered mostly to small, rural markets and carried popular programs such as the *The Lone Ranger and The Green Hornet*.[10]

The rise of radio networks gave way to a new generation of radio listeners. No longer a system of amateurs broadcasting to a puny local audience, radio became a medium to which people from all over the nation could connect and enjoy the same entertainment, share information, and exchange opinions. In 1935, two out of three homes had a radio set, and broadcasting could be heard 24 hours a day. By 1950, there were radio sets in 40 million homes, and 63 percent of Americans claimed radio as their primary source for news. This era of initial excitement and respect for the medium is often referred to as the "golden age" of radio. However, the golden age came to gradual extinction as radio became a victim of its own growth. More stations meant more competition, and a swollen market meant less revenue to go around.[11]

© Ensuper/Shutterstock.com

The Rise of FM

Although Lee de Forest criticized Edwin Armstrong for being an unoriginal thinker, Armstrong proved himself as an inventor with the creation of frequency modulation (**FM**) broadcasting. Unlike standard amplitude modulation (**AM**) broadcasting, which varies the amplitude, or power, of radio waves to create sound, FM varies the frequency of radio waves across a wide band of frequency, creating a clearer sound.[12] Still, AM radio dominated the broadcast band until the 1960s. Although the technology for the static-free sound of FM broadcasting had been around for nearly 30 years, concern for sound quality did not equal the concern for bandwidth exertion. Finally, in 1961, new Federal Communications Commission (FCC) regulations opened up more bandwidth on the FM dial. The combination of newly introduced stereo sound, the transistor radio, and high-fidelity records helped FM radio become the home for music.

As a larger audience began to tune in, FM stations became more specialized. Tailored formats such as easy listening or rock and roll started to emerge. Baby boomers, who were becoming teenagers during FM's emergence, could listen to music specifically targeted to their generation. FM radios were

---

[10] "History of Radio," http://history.sandiego.edu/GEN/recording/radio.html. Accessed July 22, 2008.

[11] Steven Schoenherr, "Golden Age of Radio 1935–50," *Recording Technology History* Accessed July 6, 2005, at http://history.sandiego.edu/gen/recording/notes.html.

[12] modulation. *Encyclopedia Britannica* at http://www.britannica.com/EBchecked/topic/387402/modulation. Accessed July 22, 2008.

showing up in people's homes, cars, and even in their hands. In the early 1960s, Japanese electronics company Sony perfected the transistor radio with the TR-63. The small, stylish, and battery-powered FM radio could be carried in a front shirt pocket. The transistor radio allowed people to stay connected to music, news, and entertainment while on the go, similar to the way MP3 players do today.

As contemporary music continued to change throughout the 1960s, station formats became more experimental. DJs started to do more than just play records; they indulged in political debates over controversial issues such as the Vietnam War, civil rights, and the draft.[13] The "free-form" format of FM stations helped shape the music and politics of the 1960s and 70s. This mini renaissance for radio faded as baby boomers aged and as FCC rule changes in the 1990s made it easier for large corporations to purchase multiple independent stations.

Talk Radio

AM radio was shedding audience in the 1970s, as the best music migrated to FM. In the 1980s, however, the talk format caught on and flourished, revitalizing AM radio. The AM frequencies that couldn't handle high-fidelity audio well were fine for transmitting the sound of the human voice. The 1987 repeal of The Fairness Doctrine (see Figure 8.1) further hastened the rise of talk radio. Previously, the Fairness Doctrine had required broadcasters to balance shows in terms of their political points of view. The idea was that the need for the public to have access to a variety of viewpoints outweighed commercial needs of broadcasters. With the law's repeal, however, political commentators such as Rush Limbaugh could bring a blend of news, entertainment, and partisan analysis to AM. Although the politically charged broadcasts of commentators such as Limbaugh often spark controversy, they also attract listeners. Talk radio is the top choice for about 15 percent of the radio audience, and it dominates AM radio today. Listeners are mostly male, but otherwise fairly diverse.[14]

Rules and Regulations

From the beginning, the radio industry has been subject to a variety of government rules and regulations. As times and technology have changed, so too have many of the rules governing the medium.

The Early Years

Signal interference arising from broadcasts airing on the same frequencies is a technical issue requiring a large-scale solution. Unless a station was granted a "clear channel" to broadcast on the same frequency across the entire country, radio stations from different locations broadcast on the same frequencies. Trouble arose when stations strayed into each other's frequencies or when their signals were too powerful. The result was static, dissonance, and disgruntled audiences.

---

[13] *Historical Dictionary of American Radio*, edited by Donald G. Godfrey and Frederic A. Leigh (Greenwood Publishing Group, 1998).

[14] "State of the News Media 2008: Audience," The Project for Excellence in Journalism, *Journalism.org*, at http://www.stateofthe-newsmedia.org/2008/narrative_radio_audience.php?cat=2&media=10. Accessed July 17, 2008.

INSIGHT

Section 315 Affects Candidates for Office

Perhaps the best-known part of the Communications Act of 1934 is Section 315, which stipulates that if a broadcast licensee permits one legally qualified candidate for public office to use a broadcasting station, equal opportunity shall be afforded to all other candidates. The licensee has no power of censorship of material broadcast under provisions of Section 315, nor does that licensee have any obligation to allow the use of the station by any candidates.

Subsequently the act was amended so that appearance by a legally qualified candidate on a bona fide newscast, news interview, documentary, or spot coverage of news events is not covered by the provisions of Section 315.

Every four years, Congress usually passes special legislation exempting presidential contests from the provisions of Section 315 so that national debates can be held without affording "equal opportunity" to fringe candidates who have no realistic chance of garnering enough votes to be serious contenders.

Section 315 applies only to appearances by the candidates themselves, not their supporters or discussions of their views or platforms. When a candidate is a movie or television star, broadcast stations refuse to run the candidate's entertainment material for the duration of the contest. For that reason, none of Ronald Reagan's old movies were shown on television during his campaigns.

In the United States, radio was a commercial affair almost from the beginning. Herbert Hoover served as Secretary of Commerce under presidents Harding and Coolidge before being elected president in 1928. He favored the allocation of radio frequencies to large corporations as a solution to the problems of interference, ownership, and funding.

In 1923, three kinds of stations were created by allocating bands of frequencies. These included high-power "clear" channels, medium-power channels, and low-power channels. The Radio Act of 1927 set up the Federal Radio Commission (FRC) with responsibility for regulating radio frequency usage. In 1928, the FRC's General Order 40 shifted most radio stations' frequencies so that 23 of the first 25 clear channels were assigned to affiliates of NBC. Formed in 1926, NBC had 48 affiliate stations, sold "sustaining" programs to affiliates, and broadcast "sponsored" programs produced by advertisers such as the American Tobacco Company.

The FRC's favoritism toward commercial interests gave rise to a radio reform movement beginning in 1930. An examination of the impact of radio and newspaper publishing on democracy in 1934 by the Hutchins Commission helped lead to the establishment of the FCC to replace the FRC. Many of the problems identified by the Hutchins Commission—such as the distortion and neglect of news coverage by large corporations and the control of news by advertising and commercial interests—remain relevant today.

The ownership caps and other rules and regulations put in place by the FCC in 1934 grew out of the Hutchins Commission recommendations. This 1934 legislation, updated by amendments, still influences policy today.[15]

The 1980s and Beyond

As a result of further technological developments in the telecommunications industry, such as the introduction of cable television, the FCC continued to update its rules and regulations over the years. In the 1980s however, President Reagan began actively moving to deregulate industries. His efforts had their greatest impact on radio with the Telecommunications Act of 1996, which President Clinton signed into law. This act includes many sections and deals with television, cable, telephony, radio, and more. Its impact on broadcast radio in the United States has been far-reaching.[16]

Before 1996, corporations were not allowed to own more than 40 radio stations. Deregulation in the 1980s and 1990s, however, relaxed these limits. Business strategists sought to make money by consolidating smaller independent companies into large corporations. By exploiting economies of scale, giant corporations produced economic gains for their stockholders.

After the deregulation in 1996, media consolidation snowballed in AM and FM radio in the United States. Today, six companies own the majority of the nation's radio broadcasting frequencies and therefore dominate broadcast radio. These *oligopolies* exert economic dominance over radio and related media enterprises such as advertising, performance venues, and recording contracts for music artists and merchandizing. Clear Channel, one of just a handful of companies that control most of broadcast radio frequencies, aggressively acquired stations when the ownership limit was relaxed, and today they control over 1,200 radio stations nationwide.[17] They dominate broadcast radio in 100 of the 112 major media markets.[18]

Increased media consolidation led to a decline in FM radio's individuality and local character. Most owners were increasing the number of ads per hour and decreasing the number of songs in their playlists. Restricted playlists with as few as 100 songs became commonplace. In 2005, on Clear Channel stations, commercials took one minute in five of typical broadcasts. Audiences began to turn away from programming that trailed in importance to advertising.

---

[15] Christopher H. Sterling, "U.S. Policy: The Communications Act of 1934" *U.S. Museum of Broadcast Communications Archives*, at http://www.museum.tv/archives/etv/U/htmlU/uspolicyc/uspolicyc.htm. Accessed July, 17, 2008.

[16] Fritz J. Messere, "U.S. Policy: Telecommunications Act of 1996," *U.S. Museum of Broadcast Communications Archives*, at http://www.museum.tv/archives/etv/U/htmlU/uspolicyt/uspolicyt.htm. Accessed July 17, 2008.

[17] Clear Channel Communications Homepage. *Clear Channel Communications, Inc.*, at http://www.clearchannel.com/. Accessed July 17, 2008.

[18] Jeff Perlstein, "Clear Channel: The Media Mammoth that Stole the Airwaves," *Corp Watch.com*, November 14, 2002, at http://www.corpwatch.org/article.php?id=4808. Accessed July 28, 2008.

Public Interest Rouses FCC

Although the Telecommunications Act of 1983 did remove many of the regulatory controls on commercial radio, the broadcast airwaves continue to belong to the public, and the FCC still retains responsibility for regulating them.

In 2003, the FCC moved to relax the rules on media ownership further. This may have been a tipping point, because public outcry calling for limits on media concentration was fierce. Nearly 3 million people contacted the FCC calling for media reform in 2003. The issue brought together groups from both the left and the right, including the National Rifle Association, Common Cause, MoveOn.org, the Traditional Values Coalition, and Mediareform.org.

The matter went to court. There was no further relaxation of ownership limits, and the FCC decided not to appeal the matter. As a result, the FCC had to rewrite its policy pertaining to serving local interests and promoting access to the airwaves by diverse groups.[19]

Developments and Issues in Contemporary Radio

The radio industry has experienced dramatic changes in the recent years. For us in the 21st century, the consumer is king of mass media. With digital media and broadband networks, consumers can get content wherever, whenever, and however they want. Broadcasting is giving way to **narrowcasting** in radio, as well as other mass media. The question isn't AM or FM, nor is it what station to tune in. Today's listener wants "me media" that provides whatever content he or she wants at the time and place demanded by the listener. Today, radio transmissions are often "narrowcast" to listeners on demand.

Satellite Radio

The rise of **satellite radio** in recent years marks an important development in the radio industry. Subscribers to satellite networks pay a monthly fee and get access to commercial-free programming, some of it directly from broadcast radio, and other stations that are only heard via satellite.

Satellite radio operators have made deals with car manufacturers so that all new vehicles have built-in satellite radio receivers, and most come with a complimentary subscription to the first several months of satellite radio.

Portable satellite radios are also available. These devices can download and play individual songs like MP3 players can, as well as play programming streams from satellite radio.

---

[19] Paul Davidson, "Looser media rules tossed," USA TODAY, June 24, 2004, at http://www.usatoday.com/money/media/2004-06-24-fcc-overturned_x.htm. Accessed July 28, 2008.

Other listeners time shift their favorite radio programs by subscribing to RSS (real simple syndication) feeds that automatically download the most recent episodes of a program to their computer and transfer the segments to their MP3 players.

In 2006, **shock-jock** Howard Stern moved to satellite radio. The move was closely watched throughout the radio business. He commanded an audience of 10 million with his morning show on FM radio. Following crackdowns on profanity over the airwaves from regulators and the threat of fines or censorship from his employer, Stern decided to make a decisive move off the public airwaves and onto subscription radio broadcast via satellite. With his move from FM broadcast radio to satellite radio, his new employer SIRIUS saw its subscription base rise above the 3-million mark.

In the past several years, satellite radio subscriptions have continued to increase. After the 2008 merger of SIRIUS and XM, the combined company had over 18 million subscribers.[20] That same year, a study conducted by the digital technology research firm Parks Associates predicted that by 2012, satellite radio subscribers would increase to 39 million.[21]

Public Broadcasting

More than 26 million people listen to Corporation for Public Broadcasting (CPB) stations and support news shows like "All Things Considered," long-form radio entertainments like "This American Life," and a variety of musical offerings on public radio stations through pledges. The audience for CPB had remained steady, but has declined slightly between 2004 and 2005. Competition for listener attention is fierce, and the hours per day that individuals can engage with mass media can't expand indefinitely. It may be inevitable that as more varied devices are available, the number of listeners in the audience for any one device may decrease.

Internet Radio and Podcasting

Innovative radio broadcasting stations of all kinds are presenting supplemental content via downloads for listeners who have special interests. News radio 780, one of the only all-news AM radio stations still broadcasting, has put press conferences and uncut interviews on its Web site where listeners can listen, download, or subscribe to news of their choice. The Internet has allowed the station, and others like it, to reach audiences far beyond their over-the-airwaves reach.[22]

---

[20] "SIRIUS XM Radio Provides Merger Update," September 9, 2008, SIRIUS Satellite Radio News Releases, http://investor.sirius.com/ReleaseDetail.cfm?ReleaseID=333252&cat=&newsroom. Accessed September 12, 2008.

[21] "Almost 70 Million Consumers Will Have Either an HD or a Satellite Radio by 2012," *Reuters*, January 8, 2008, at http://www.reuters.com/article/pressRelease/idUS204500+08-Jan-2008+PRN20080108. Accessed September 12, 2008.

[22] WBBM 780 homepage, at http://www.wbbm780.com/. Accessed July 18, 2008.

Nowadays, even individuals with computers and inexpensive software can provide competition for listeners. It is simple to create online music or talk programming. Digital audio, such as **podcasts** (digital recordings of radio broadcasts or similar programs made available on the Internet),[23] can be uploaded to the Internet and downloaded for listening on computers or MP3 players. This is yet another form of competition for the attention of listeners.

Revenues and Ratings

Radio revenues are dependent on advertising. But how do radio stations know how large the audience is for various programs, and how do advertisers verify that the claims made by station owners are correct? They turn to Arbitron Inc., an "international media and marketing research firm serving radio broadcasters, radio networks, cable companies, advertisers, advertising agencies, outdoor advertising companies, and the online radio industry in the United States, Mexico, and Europe."[24] In a joint venture with VNU, Inc., Arbitron provides media and marketing research services including market **ratings** to broadcasters and advertisers.

Radio shows are programmed around a set of genres. A **genre** is a style. Like any other fashion or any other style, music genres go in and out of popularity with audiences. Arbitron also provides data on audience demographics for different genres and for various **dayparts**.

The amount of money that broadcasters can use for advertising time is tied to ratings. Arbitron and similar companies provide ratings information for major markets. These resources provide an interesting way to learn about radio and are essential for those who intend to pursue a career in radio.

Responding to Competition

With satellite radio, Internet radio, and podcasting bursting onto the media scene, there are suddenly many more alternatives for the listening audience.[25] To stay in business, the radio industry looked to new formats as a way to be competitive in a world in which competition comes from the Internet, MP3 devices, cell phones, or anyone who has a computer with a microphone and Internet connection.

In 2002, an alternative format called "Jack" was trademarked in Canada. With the slogan "Playing what we want," "Jack" stations typically have a playlist of 1,000 songs. They are classified as variety or adult hit by radio research companies.[26]

By 2004, "Jack," "Bob," and even "Dave" radio, based loosely on the "random shuffle" concept, were taking over American airwaves. Several of these alternatives to the Clear Channel **cyberjock**

---

[23] podcast. *New Oxford American Dictionary* at http://www.bbn.com/utility/glossary/p. Accessed July 18, 2008.

[24] About Arbitron. *Arbitron, Inc.*, at http://www.arbitron.com/about/home.htm. Accessed July 18, 2008.

[25] Podcasting News homepage at http://www.podcastingnews.com/. Accessed July 17, 2008.

[26] Lou Pickney, "Variety Hits-History of the Format," at http://www.varietyhits.com/variety/history.shtml. Accessed July 17, 2008.

model began to advertise fewer commercials per hour. Although they denied they were responding to the competition, Clear Channel moved to reduce the number of commercials on its stations in 2005.

In addition, companies such as Clear Channel responded to mounting competition by seeking to cut expenses by achieving economies of scale. Clear Channel is a leader in using **voice-tracking software** to customize their programs to make it sound as if the DJs are actually local residents. With digital equipment, local traffic and weather information can be interlaced into a broadcast from a single point, like Clear Channel's San Antonio studios. When the broadcast goes out across the airwaves, listeners think the DJ is local because of the customized traffic and weather segments. The use of cyberjocks cuts the cost of producing radio shows, but at the same time it can rob the airwaves of diverse voices and local perspectives.[27] Other changes evolve from the consolidation of radio ownership.

Beyond National Boundaries: Global Radio

Although radio is an old medium, it is still one of the most vital forms of communication outside the United States. There are about 44,000 radio stations in the world according to the CIA World Factbook, and only one in three of these is in the United States.

The BBC

In the United States, the airwaves were largely put into commercial use. In the United Kingdom, the British Broadcasting Corporation BBC, or the "Beeb" as it is called by millions of listeners, was established with public support and a royal charter in the 1927.[28] The effort to keep the BBC independent of commercial and government influence was led by John Reith, the BBC's first general manager, who believed in "an independent British broadcaster able to educate, inform, and entertain the whole nation, free from political interference and commercial pressure."[29]

In 1937, the BBC established a network designed for its empire and the rest of the world, which is called the World Service today. The BBC was broadcasting television by 1936, but had to suspend television broadcasts during World War II.

The coronation of Queen Elizabeth in 1952 drew the largest audience in BBC history and set the stage for television to become the premiere broadcast format. BBC programming was solid, but competition from Independent Television (ITV) captured many viewers during the 1950s and 1960s. In the 1970s, the BBC linked with the Open University to create the University of

[27] Anna Wilde Matthews, "Clear Channel Uses High-Tech Gear To Perfect the Art of Sounding Local," February 25, 2002, at http://stayfreemagazine.org/public/clearchannel.html.

[28] BBC Radio homepage, at http://www.bbc.co.uk/radio/i/. Accessed July 17, 2008.

[29] "The BBC Story: History of the BBC, the 1920s" at http://www.bbc.co.uk/heritage/story/index.shtml. Accessed July 17, 2008.

the Airwaves. In the 1980s, Margaret Thatcher's government cut back funding, but the BBC streamlined and in the 1990s launched the highly respected BBC Online. By January 2005, BBC Internet radio listenership had jumped 70 percent from the previous year, with a total of over 7 million listeners.[30] Today, the BBC Web site and Internet radio broadcasts continue to gain an audience.

Shortwave Broadcasts

Many radio stations that are widely accessible are shortwave broadcasts. Shortwave signals can be aimed at the earth's ionosphere. When they strike the ionosphere, they bounce back to Earth and can skip from one part of the globe. This is called the skywave effect, and it helps the waves continue or propagate over long distances.

Solar emissions can affect the reflectivity of the earth's atmosphere, but if conditions are right, a shortwave radio in the United States can pick up stations from Africa, Asia, Australia, and East Asia.

Avid shortwave listeners are known as Dxers from the old telegraph term meaning "Distance." They search out broadcasts that fall into three main categories: broadcast, clandestine, and amateur or utility. Broadcasts consist of music, cultural, religious, and political programming. "Pirate" and other unlicensed radio stations, including some anti-government broadcast material, fall into the clandestine category. Utility broadcasts include weather information, world clocks, Morse code, and radio teletype.

[30] "BBC internet radio users up 70% on 2004," *BBC Press Office*, February 28, 2005, at http://www.bbc.co.uk/pressoffice/pressreleases/stories/2005/02_february/28/radio.shtml.

Name ______________________ Course ______________ Date ________________

Chapter Questions

1. From the beginning, the radio industry has been subject to numerous federal rules and regulations. After reading this section, do you think that radio should be governmentally controlled? Why do you feel this way?

2. What effect has satellite radio, podcasts, and streaming radio had on traditional AM/FM radio? What can traditional radio give an audience that other sources cannot? On the other hand, how are satellite and streaming radio more beneficial than AM/FM radio?

THUNDERSTUCK

Erik Larson

That completed electric circuits emit invisible waves had been demonstrated by scientists such as Heinrich Hertz and Oliver Lodge by the 1890s, but no one had yet found a practical use for these electromagnetic waves. Lodge, an Englishman, might have found them but was distracted by an interest in the occult. It was left to an Italian "tinkerer" with little formal education, Gugliemo Marconi, to formulate an historic advancement that became radio. In his novel, Thunderstruck, noted author Erik Larson details Marconi's race to develop practical radio before anyone else. In this excerpt Marconi has read about electromagnetic waves in science magazines and works frenetically to create components from scratch that will defy popular physics and make him wealthier than he can imagine.

© muratart/Shutterstock.com

In the beginning, in the invisible realm where electromagnetic energy traveled, there was emptiness. Such energy did exist, of course, and traveled in the form of waves launched from the sun or by lightning or any random spark, but these emanations rocketed past without meaning or purpose, at the speed of light. When men first encountered sparks, as when a lightning bolt incinerated their neighbors, they had no idea of their nature or cause, only that they arrived with a violence unlike anything else in the world. Historians often place humankind's initial awareness of the distinct character of electrical phenomena in ancient Greece, with a gentleman named Thales, who discovered that by rubbing amber he could attract to it small bits of things, like beard hair and lint. The Greek word for amber was *elektron*.

As men developed a scientific outlook, they created devices that allowed them to generate their own sparks. These were electrostatic machines that involved the rubbing of one substance against another, either manually or through the use of a turning mechanism, until enough electrostatic charge—static electricity—built up within the machine to produce a healthy spark or, in the jargon of electrical engineers, a disruptive discharge. Initially scientists were pleased just to be able to launch a spark, as when Isaac Newton did it in 1643, but the technology quickly improved and, in 1730, enabled one Stephen Gray to devise an experiment that for sheer inventive panache outstripped anything that had come before. He clothed a boy in heavy garments until his body was thoroughly insulated but left the boy's hands, head, and feer naked. Using nonconducting silk strings, he hung the boy in the air, then touched an electrified glass tube to his naked foot, thus causing a spark to rocket from his nose.

The study of electricity got a big boost in 1745 with the invention of the Leyden jar, the first device capable of storing and amplifying static electricity. It was invented nearly simultaneously in Germany and in Leyden, the Netherlands, by two men whose names did not readily trip from the tongue: Ewald Jürgen von Kleist and Pieter van Musschenbroek. A French scientist, the Abbé Nollet, simplified things by dubbing the invention the Leyden phial, although for a time a few proprietary Germans persisted in calling it a von Kleist bottle. In its best-known iteration, the Leyden jar consisted of a glass container with coatings of foil on the inside and outside. A friction machine was used to charge, or fill, the jar with electricity. When a wire was used to link both coatings, the jar released its energy in the form of a powerful spark. In the interests of science Abbé Nollet went on to deploy the jar to make large groups of people do strange things, as when he invited two hundred monks to hold hands and then discharged a Leyden jar into the first man, causing an abrupt and furious flapping of robes.

Naturally a competition got under way to see who could launch the longest and most powerful spark. One researcher, George Richman, a Swede living in Russia, took a disastrous lead in 1753 when, in the midst of an attempt to harness lightning to charge an electrostatic device, a huge spark leaped from the apparatus to his head, making him the first scientist to die by electrocution. In 1850 Heinrich D. Ruhmkorff perfected a means of wrapping wire around an iron core and then rewrapping the assembly with more wire to produce an "induction coil" that made the creation of powerful sparks simple and reliable—and incidentally set mankind on the path toward producing the first automotive ignition coil. A few years later researchers in England fashioned a powerful Ruhmkorff coil that they then used to fire off a spark fortytwo inches long. In 1880 John Trowbridge of Harvard launched a sevenfooter.

Along the way scientists began to suspect that the sudden brilliance of sparks might mask deeper secrets. In 1842 Joseph Henry, a Princeton professor who later became the first director of the

Smithsonian Institution, speculated that a spark might not be a onetime burst of energy but in fact a rapid series of discharges, or oscillations. Other scientists came to the same conclusion and in 1859 one of them Berend Fedderson, proved it beyond doubt by capturing the phenomenon in photographs.

But it was James Clerk Maxwell who really shook things up. In 1873 in his *A Treatise on Electricity and Magnetism* he proposed that such oscillations produced invisible electromagnetic waves, whose properties he described in a series of famous equations. He also argued that these waves were much like light and traveled through the same medium, the mysterious invisible realm known to physicists of the day as ether. No one yet had managed to capture a sample of ether, but this did not stop Maxwell from calculating its relative density. He came up with the handy estimate that it had 936/1,000, 000,000,000,000,000,000ths the density of water. In 1886 Heinrich Hertz proved the existence of such waves through laboratory experiments and found also that they traveled at the speed of light.

Meanwhile other scientists had discovered an odd phenomenon in which a spark appeared to alter the conducting properties of metal filings. One of them, Edouard Branly of France, inserted filings into glass tubes to better demonstrate the effect and discovered that simply by tapping the tubes he could return the filings to their nonconducting state. He published his findings in 1891 but made no mention of using his invention to detect electromagnetic waves, though his choice of name for his device was prophetic. He called it a radio-conductor. At first his work was ignored, until Oliver Lodge and his peers began to speculate that maybe Hertz's waves were what caused the filings to become conductive. Lodge devised an improved version of the Branly tube, his "coherer," the instrument he unveiled at the Royal Institution.

Lodge's own statements about his lecture reveal that he did not think of Hertzian waves as being useful; certainly the idea of harnessing them for communication never occurred to him. He believed them incapable of traveling far—he declared half a mile as the likely limit. It remained the case that as of the summer of 1894 no means existed for communicating without wires over distances beyond the reach of sight. This made for lonely times in the many places where wires did not reach, but nowhere was this absence felt more acutely than on the open sea, a fact of life that is hard to appreciate for later generations accustomed to the immediate world-grasp afforded by shortwave radio and cellular telephone.

The completeness of this estrangement from the affairs of land came home keenly to Winston Churchill in 1899 on the eve of the Boer War, when as a young war correspondent he sailed for Cape Town with the commander of Britain's forces aboard the warship *Dunottar Castle*. He wrote, "Whilst the issues of peace and war seemed to hang in their last flickering balance, and before a single irrevocable shot had been fired, we steamed off into July storms. There was, of course, no wireless at sea in those days, and, therefore, at this most exciting moment the Commander-in-Chief of the British forces dropped completely out of the world. After four days at sea, the ship called at Medeira where there was no news. Twelve days passed in silence and only when the ship was two days from Cape Town was another ship sighted coming from the 'land of knowledge' and bearing vital news. Signals"—visual signals—"were made to the steamer, a tramp, asking for news, upon which she altered course to pass within a hundred yards of the *Dunottar Castle*, and held up a blackboard bearing the words, 'Three battles. Penn Symonds killed.' Then she steamed on her way, and the Commander-in-Chief, whose troops had been in action without his knowledge, was left to meditate upon this very cryptic message."

Back from the Alps, Marconi immediately set to work devising equipment to transform his idea into reality, with nothing to guide him but an inner conviction that his vision could be achieved. His mother recognized that something had changed. Marconi's tinkering had attained focus. She saw too that now he needed a formal space dedicated to his experiments, though she had only a vague sense of what it was that he hoped to achieve. She persuaded her husband to allow Marconi to turn a portion of the villa's third-floor attic into a laboratory. Where once Marconi's ancestors had raised silkworms, now he would coils of wire and fashioned Leyden jars that snapped blue with electrical energy.

On hot days the attic turned into a Sahara of stillness. Marconi grew thin, his complexion paler than usual. His mother became concerned. She left trays of food on the landing outside the attic door. Marconi's father, Giuseppe, grew increasingly unhappy about Marconi's obsession and its jarring effect on family routine. He sought to reassert control by crimping his already scant financial support for his son's experiments. "Giuseppe was punishing Guglielmo in every way he knew," wrote Degna. "Characteristically he considered money a powerful weapon." At one point Marconi sold a pair of shoes to raise money to buy wire and batteries, but this clearly was a symbolic act meant to garner sympathy from his mother, for he had plenty of shoes to spare.

In his attic laboratory Marconi found himself at war with the physical world. It simply was not behaving as he believed it should. From his reading, Marconi knew the basic character of the apparatus he would need to build. A Leyden jar or Ruhmkorff coil could generate the required spark. For a receiver, Marconi built a coherer of the kind Branly had devised and that Lodge had improved, and he connected it to a galvanometer, a device that registered the presence of an electrical current.

But Marconi found himself stymied. He could generate the spark easily but could not cause a response in his coherer. He tinkered. He tried a shorter tube than that deployed by Lodge, and he experimented with different sizes and combinations of filings. At last he got a response, but the process proved fickle. The coherer "would act at thirty feet from the transmitter," Marconi wrote, but "at other times it would not act even when brought as close as three or four feet."

It was maddening. He grew thinner, paler, but kept at it. "I did not lose courage," he wrote. But according to Degna, "he did lose his youth" and took on a taciturnity that, by her account, would forever color his demeanor.

He wanted distance. He knew that if his telegraphy without wires was ever to become a viable means of communication, he would need to be able to send signals hundreds of miles. Yet here in his attic laboratory he sometimes could not detect waves even an arm's length from the spark. Moreover, established theory held that transmitting over truly long distances, over the horizon, simply was not possible. The true scholar-physicists, like Lodge, had concluded that waves must travel in the same manner as light, meaning that even if signals could be propelled for hundreds of miles, they would continue in a straight line at the speed of light and abandon the curving surface of the earth.

Another man might have decided the physicists were right—that long-range communication was impossible. But Marconi saw no limits. He fell back on trial and error, at a level of intensity that verged on obsession. It set a pattern for how he would pursue his quest over the next decade. Theoreticians devised equations to explain phenomena; Marconi cut wire, coiled it, snaked it, built apparatus, and flushed it with power to see what would happen, a seemingly mindless process but one governed by the certainty that he was correct. He became convinced, for example, that the composition of the metal filings in the coherer was crucial to its performance. He bought or scavenged metals of all kinds and used a chisel to scrape loose filings of differing sizes, then picked through the filings to achieve uniformity. He tried nickel, copper, silver, iron, brass, and zinc, in different amounts and combinations. He inserted each new mixture into a fragile glass tube, added a plug of silver at each end, then sealed the apparatus and placed it within his receiving circuit.

He tested each mixture repeatedly. No instrument existed to monitor the strength or character of the signals he launched into space. Instead, he gauged performance by instinct and accident. He did this for days and weeks on end. He tried as many as four hundred variations before settling on what he believed to be the best possible combination for his coherer: a fine dust that was 95 percent nickel and 5 percent silver, with a trace of mercury.

At first he tried to use his transmitter to ring a bell at the far side of his laboratory. Sometimes it worked, sometimes not. He blamed the Branly-style coherer, calling it "far too erratic and unreliable" to be practical. Between each use he had to tap it with his finger to return the filings to their nonconducting state. He tried shrinking the size of the tube. He emptied thermometers, heated the glass, and shaped it. He moved the silver plugs within the tube closer and closer together to reduce the expanse of filings through which current would have to flow, until the entire coherer was about an inch and a half long and the width of a tenpenny nail. He once stated that it took him a thousand hours to build a single coherer. As a future colleague would put it, he possessed "the power of continuous work."

Guglielmo Marconi with an early version of his radio

Marconi's obsession with distance deepened. He moved the bell to the next room and discovered how readily the waves passed through obstacles. As he worked, a fear grew within him, almost a terror, that one day he would awaken to discover that someone else had achieved his goal first. He understood that as research into electromagnetic waves advanced, some other scientist or inventor or engineer might suddenly envision what he had envisioned.

And in fact he was right to be concerned. Scientists around the world were conducting experiments with electromagnetic waves, though they still focused on their optical qualities. Lodged had come closest, but inexplicably had not continued his research.

GunFire

Each moment mattered. Marconi shrank the size of his coherer until the space containing the filings was little more than a slit between the two silver plugs. He tried heating the glass tube just before he sealed it so that once the air within cooled and contracted to room temperature it would create a partial vacuum. This by itself caused a marked improvement in the coherer's sensitivity.

A persistent irritant was the need to tap the coherer to return it to a state where it once again could respond to passing waves. No telegraph system could survive such a laborious and imprecise procedure.

Marconi devised a clapper like that from a bell and inserted it into the receiving circuit. "Every time I sent a train of electric waves," Marconi wrote, "the clapper touched the tube and so restored the detector at once to its pristine state of sensibility."

He took his experiments outdoors. He managed to send the Morse letter S, three dots, to a receiver on the lawn in front of the villa. With additional tinkering and tweaking to improve the efficiency of his circuits, he increased his range to several hundred yards. He continued trying new adjustments but could send no farther.

One day, by chance on intuition, Marconi elevated one of the wires of his transmitter on a tall pole, thus creating an antenna longer than anything he previously had constructed. No theory existed that even hinted such a move might be useful. It was simply something he had not yet done and that was therefore worth trying. As it happens, he had stumbled on a means of dramatically increasing the wavelength of the signals he was sending, thus boosting their ability to travel long distances and sweep around obstacles.

"That was when I first saw a great new way open before me," Marconi said later. "Not a triumph. Triumph was far distant. But I understood in that moment that I was on a good road. My invention had taken life. I had made an important discovery."

It was a "practician's" discovery. He had so little grasp of the underlying physics that later he would contend that the waves he now harnessed were not Hertzian waves at all but something different and previously unidentified.

Enlisting the help of his older brother, Alfonso, and some of the estate's workers, he experimented now with different heights for his antennas and different configurations. He grounded each by embedding a copper plate in the earth. At the top he attached a cube or cylinder of tin. He put Alfonso in charge of the receiver and had him carry it into the fields in front of the house.

He began to see a pattern. Each increase in the height of his antenna seemed to bring with it an increase in distance that was proportionately far greater. A six-foot antenna allowed him to send a signal sixty feet. With a twelve-foot antenna, he sent it three hundred feet. This relationship seemed to have the force of physical law, though at this point even he could not have imagined the extremes to which he would go to test it.

Eventually Marconi sent Alfonso so far out, he had to equip him with a tall pole topped with a handkerchief, which Alfonso waved upon receipt of a signal.

The gain in distance was encouraging. "But," Marconi said, "I knew my invention would have no importance unless it could make communication possible across natural obstacles like hills and mountains."

Now it was September 1895, and the moment had come for the most important test thus far.

He sat at the window of his attic laboratory and watched as his brother and two workers, a farmer named Mignani and a carpenter named Vornelli, set off across the sun-blasted field in front of the house. The carpenter and the farmer carried a receiver and a tall antenna. Alfonso carried a shotgun.

The plan called for the men to climb a distant hill, the Celestine Hill, and continue down the opposite flank until completely out of sight of the house, at which point Marconi was to transmit a signal. The distance was greater than anything he had yet attempted—about fifteen hundred yards—but far more important was the fact that it would be his first try at sending a signal to a receiver out of sight and thus beyond the reach of any existing optical means of communication. If Alfonso received the signal, he was to fire his shotgun.

The attic was hot, as always. Bees snapped past at high velocity and confettied the banks of flowers below. In a nearby grove silver-gray trees stood stippled with olives.

Slowly the figures in the field shrank with distance and began climbing the Celestine Hill. They continued walking and eventually disappeared over its brow, into a haze of gold.

The house was silent, the air hot and still. Marconi pressed the key on his transmitter.

An instant later a gunshot echoed through the sun-blazed air.

At that moment the world changed, though a good deal of time and turmoil would have to pass before anyone was able to appreciate the true meaning of what just had occurred.

Name ______________________ **Course** ______________ **Date** ________________

Chapter Questions

1. Marconi was not a trained scientist and did not know a great deal about the "Hertz" waves and the laws of Physics as they were understood in the 1890s. Why might this have been a good thing?

2. In the last portion of this excerpt Marconi was able to send electromagnetic waves to a spot out of sight and over a hill. "At that moment the world changed…" writes the author. Do you believe that is an exaggeration or a sound statement?

FAKE WAR ON RADIO SPREADS PANIC OVER U.S.

George Dixon

"War of the Worlds" is a radio drama that was directed and narrated by Orson Welles on October 30, 1938, and based on the science fiction novel by H.G. Wells. The broadcast, which aired as a fictionalized news bulletin, realistically portrays violent extraterrestrials attacking parts of the United States and killing everyone in their path. In an atmosphere of unease and fear during the buildup to World War II, many listeners panicked, believing the events and news broadcast to be real. Because of the public's reaction, "War of the Worlds" is often described as the most famous and impactful radio broadcast in history. This New York Daily News article was written the day after the broadcast aired and details the panicked reactions of people. Over 12,500 articles, like this one, were published in the days following reassuring people that the radio show was fictional and easing their anxiety.

A radio dramatization of H.G. Wells' "War of the Worlds" - which thousands of people misunderstood as a news broadcast of a current catastrophe in New Jersey - created almost unbelievable scenes of terror in New York, New Jersey, the South and as far west as San Francisco between 8 and 9 o'clock last night.

The panic started when an announcer suddenly interrupted the program of a dance orchestra - which was part of the dramatization - to "flash" an imaginary bulletin that a mysterious "meteor" had struck New Jersey, lighting the heavens for miles around.

A few seconds later, the announcer "flashed" the tidings that weird monsters were swarming out of the mass of metal - which was not a meteor but a tube-like car from Mars - and were destroying hundreds of people with death-ray guns.

Thousands flee

Without waiting for further details, thousands of listeners rushed from their homes in New York and New Jersey, many with towels across their faces to protect themselves from "gas" which the invader was supposed to be spewing forth.

Simultaneously, thousands more in states that stretched west to California and south to the Gulf of Mexico rushed to their telephones to inquire of newspapers, the police, switchboard operators, and electric companies what they should do to protect themselves.

The "space cartridge" was supposed to have struck at Grover's Mills, an actual town near Princeton. Names of well-known highways were used in describing the advance of the monsters. The voice had warned them to "pack up and move north because the machines are coming from Mars." The dramatization of Wells' novel had featured a fictitious speech from "The Governor of New Jersey," assuring the public that the National Guard has been mobilized to fight the "Martian monsters" and the Harlem residents had confused the mythical "Governor" with the President.

© breadkermaximus/Shutterstock.com

Churches in both New York and New Jersey were filled suddenly with persons seeking protection, and who found them, providentially as they thought, open.

At St. Michael's Hospital, in Newark, fifteen persons were treated for shock.

In New York, police and fire departments and the newspapers were swamped with telephone calls from people, apparently frightened half out of their wits.

The telephone company also was deluged. The thing finally assumed such serious proportions that the Colombia Broadcasting System put bulletins on the air explaining that the "meteor" broadcast was part of a play and that nothing untoward had happened.

The broadcasting company added that the whole thing had been somewhat in the nature of a Halloween prank. The program, which came over station WABC from 8 to 9 P.M. was presented by Orson Welles' "Mercury Theatre of the Air."

"The War of the World's was a typical H.G. Wells shocker. It described the bombardments of England by huge "space capsules" carrying warriors from Mars. These inhuman gigantic warriors laid waste to England and killed hundreds of thousands of people. Finally, they were killed by germs and infections - because they came from a planet which had no disease, and thus were susceptible to every disease. In Wells' book there was no mention of the United States.

In its dramatization last night the radio station changed the locale to America.

Within a couple of minutes of the first death and destruction bulletin the telephone calls began pouring in. Many of the callers seemed on the point of hysteria. One woman said she has relatives in the "stricken" section of New Jersey and wanted to know if their names were on the casualty lists.

Doctors offer help

The New York City Department of Health was among the first to call The News. The department wanted to know what assistance it could lend to the maimed and dying.

Hundreds of physicians and nurses were among the callers. Many of them said they were prepared to rush at once into the devastated area to aid in caring for victims.

Scores of motorists traveling through Jersey heard the broadcast and immediately detoured so as not to pass through the supposedly doomed region. Police in many small Jersey towns and villages called State Police Headquarters to offer assistance.

At Princeton University two members of the geology faculty, equipped with flashlights and hammers, started for Grover's Mills, two miles away, where the meteor supposedly fell. Dozes of cares were driven to the hamlet my curious motorists. A score of university students were phoned by their parent, and told to come home.

An anonymous and somewhat hysterical girl phoned the Princeton Press Club from Grover's Mills and said:

"You can't imagine the horror of it! It's hell!"

A man came into the club and said he saw the meteor strike the earth and witnessed animals jumping from the alien body.

In Watchung, N.J., an excited policeman on desk duty - notified by horrified citizens that a meteor struck somewhere nearby, sent squad cars out to look for injured.

Pleas of "What can we do?" Where can we go to save ourselves?" flooded New Jersey police switchboards from Hoboken to Cape May. In Newark alone two patrolmen handled more than 2,000 calls from hysterical persons terrified by the fake news bulletin.

Harassed Newark police, trying to reassure thousands of panicky citizens, received a call about a gas explosion in a six-family house at 145 Hedden Terrace. Emergency trucks raced to the scene to find that more than thirty people, occupants of the house, were on the street, holding their clothes and bedding.

Motorists warned

In Irvington, N. J. hundreds of motorists who hear the announcement of the meteor and the gas attack shouted warnings to each other in the streets.

"Drive like hell into the country; we're being bombed by enemies", drivers shouted to one another. Motorcycle police, astounded by the sudden bursts of speed by motorists, rushed to call boxes to inquire from headquarters about the supposed raids.

In the Sacred Heart Church in Elizabeth, priests were amazed by a sudden influx of panicky persons who rushed inside, fell on their knees and began to pray.

Meanwhile, City Manager Pail Morton ofTrenton announced that he would ask an investigation of the program by the Federal Communication Commission.

State police at Morristown said that dozens of calls from irate radio listeners demanded that authorities obtain indictments against broadcast company officials.

The announcement that 7,000 National Guard members were being mobilized to defend New Jersey from the invaders, caused hundreds of guardsmen to swamp the 113th Regiment armory and the 102nd cavalry armory with calls and queries on where to report for duty.

A motorist parked near the Lodi Theatre in Orange, N. J. tuned in to the program, listened for a moment, and then ran breathlessly into the movie house.

"The state is being invaded," he screamed. "This place is going to be blown up!"

A minute later the theater was empty.

When Manuel Priola heard the broadcast in his bar at 433 Valley Road, Orange, he closed the cash register and announced to his customers:

"You folks can go where you like. I'm closing up this place and going home." And he did.

Evening services at the First Baptist Church in Caldwell, N. J., were well under way when a frantic parishioner dashed in and yelled that a meteor had fallen near by and the whole countryside was threatened. The Rev. Thomas Thomas calmed his flock and called them to pray for deliverance.

He's getting out

A terrified motorist asked Patrolman Lawrence Treger the way to Route 24.

"All creation's busted loose" he yelled, "and I'm getting out of Jersey."

Panic swept one apartment house in Greenwich Village, largely occupied by Italian families, after tenants caught scraps of the broadcast.

The super-realism of the drama sent Caroline Cantlon, an actress with a WPA Gilbert and Sullivan unit, to Polyclinic Hospital last night. Sitting in her room at the Markwell Hotel on 49th St. between Broadway and Eighth Ave., Miss Cantlon turned on her radio. She heard and announcement of smoke in Times Square. She rushed out into the hallway and down the stairs. She stumbled and fell, fracturing her arm at the wrist

How it all began

The broadcast began with the usual announcement: "The Columbia Broadcasting System presents Orson Welles and his Mercury Theatre Players in a dramatization of H. G. Wells' novel, "The War of the Worlds."

Listeners heard a dance band playing languid Spanish numbers in the "Park Astoria Hotel." The music was interrupted suddenly by a flash. It was announced that a professor in a university observatory in the Southwest had noticed a series of gas explosions on the planet Mars.

The scene switched to Princeton where an astronomer was trying to explain the phenomenon in an interview.

There was another flash. A meteor had just struck Grover's Mills, near Princeton.

The professor and an announcer rushed out to inspect the meteor. On their arrival, they reported to the radio listeners that it wasn't a meteor at all - it was a giant tube of a metal that had no earthly origin.

Then the announcer cut in excitedly:

"Just a minute. Something's happening! Ladies and gentlemen, this is terrific! The end of this things' beginning to come off. The top is beginning to rotate like a screw! The thing must be hollow!"

Then voices could be heard as other fictitious spectators grew alarmed.

"Look! The darn thing's un-screwing!"

"Keep back there. Keep back, I tell you!"

"Maybe their's men in it, trying to escape?"

"It's red hot; they'll burn to a cinder!"

"Keep back there. Keep those idiots back!"

There was the clanking sound of falling metal and then more voices:

"She's off. The top's loose, Look out there! Stand back!"

Suddenly the monsters began crawling out - huge things impervious to bullets or any other missiles. It soon became apparent that they carried some sort of fire-arms which turned out to be death rat machines. These were trained on the crowd and 200 spectators died instantly.

Martial law was declared by the Governor of New Jersey and there were bulletins from State Militia headquarters in the vicinity of the spot where the Martians were in control.

Through the drone of airplane motors could be heard the radio reports of Army pilots to their headquarters:

"...One machine partially crippled. Believed hit by shell from Army gun in Watchung Mountains. Guns now appear silent. A heavy black fog hanging over the earth...of extreme density, nature unknown..."

"...Enemy now turns east, crossing Passaic River into Jersey marshes. One of the gigantic creatures is straddling the Pulaski Skyway. Evident objective is New York City. They're pushing down a high tension power station....

"The machines are close together now and we're ready to attack. A thousands yards and we'll be over the first one!...Eight hundred...seven hundred...there they go! A giant arm is raised...there's a green flash...they're spraying us with flame! Two thousand feet. Engines are giving out. No chance to release bomb. Only one thing left to do...drop on them, plane and all. We're diving on the first one...mow the engine's gone...eight hundred feet"

Then "Operator N. 3" cut in form Newark with a warning:

"Poisonous black smoke pouring in from Jersey marshes. Gas masks useless. Urge population to move into open spaces. Automobiles use routes 7, 23, 24. Avoid congested areas..."

Then from "the roof of the broadcasting building in Madison Ave." came another bulletin:

"The bells you hear ringing are to warn the people to evacuate the city as the Martians approach. It is estimated that in the last two hours 3,000,000 people have moved out along the roads to the north. The Hutchinson River Parkway is still kept open for motor traffic. Avoid bridges to Long Island - hopelessly jammed. All communication with Jersey shores closed ten minutes ago. Our army wiped out. This may be the last broadcast...we will stay here to the end..."

From the street below came the sound of voices rising in a hymn. The announcer continued:

"I've just been handed a bulletin. Cylinders from Mars are falling all over the country. One outside Buffalo - another in Chicago - St. Louis..."

The announcer kept on. One of the monsters was approaching, "his head even with the skyscrapers." People were dripping like flies as the poison gas spread. Now the monster was crossing Sixth Ave. Now Fifth Ave. Not it was only fifty feet away..."

Just as the whole world was toppling into oblivion, the real station announcer cut in with word that the gasping audience had been listening to a dramatization of H. G. Wells' book.

War's Over: How U.S. Met Mars

The radio's "end of the world" as some listeners understood it, produces repercussions throughout the United States. Samples, as reported by the Associated Press, follow:

- Woman Tries Suicide: Pittsburgh - A man returned home in the midst of the broadcast and found his wife a bottle of poison in her hand screaming: "I'd rather die this way than like that."
- Man Wants to Fight Mars: San Francisco - An offer to volunteer in stopping an invasion from Mars came among hundreds of telephone inquiries to police and newspapers during the radio dramatization of H.G. Wells' story. One excited man called Oakland police and shouted: "My God! Where can I volunteer my services? We've got to stop this awful thing!"
- Church Lets Out: Indianapolis - A woman ran into a church screaming: "New York destroyed; it's just the end of the world. You might as well go home to die. I just heard it on the radio." Services were dismissed immediately.
- College Boys Faint: Brevard, N.C. - Five Brevard College students fainted and panic gripped the campus for a half hour with many students fighting for telephones to inform their parents to come and get them.
- It's a Massacre: Providence, R. I. - Weeping and hysterical women swamped the switchboard of the Providence Journal for details of the "massacre." The electric company received scores of calls urging it to turn off all lights so that the city would be safe from the "enemy."
- She Sees "the Fire": Boston - One woman declared she could "see the fire and told the Boston Globe she and many others in her neighborhood were "getting out of here"
- "Where is it Safe?": Kansas City - One telephone informant said he had loaded all his children into his car, had filled it with gasoline, and was goings somewhere. "Where is it safe?" he wanted to know. The Associated Press bureau received queries on the "meteors" from Los Angeles, Salt Lake City, Beaumont, Texas, and St. Joseph, Mo.
- Prayers in Richmond: Richmond, Va - The Times-Dispatch reported some of its telephone calls came from persons who said they were praying.

Senator Maps Bill to Censor Air Waves

Des Moines. Oct. 30 (AP) - Senator Clyde L. Herring (Dem', Iowa) said tonight he planned to introduce a bill in the next session of Congress "controlling just such abuses as was heard over the radio tonight." He said the bill would propose a censor-ship board to which all radio programs must be submitted.

Frank R. McNinch. Chairman of the Federal Communications Commission in Washington said that an investigation would be held at once by the FCC. He would not predict what action might be taken, but said a through probe would be made.

Wells....Welles

H.G. Wells, whose imagination has run riot for forty years, found it rather difficult to realize that anyone could have become alarmed over a broadcast of his "War of the World."

Reached by The News by trans-Atlantic telephone in London early this morning, the 72-year-old British author listened with interest to an account of what happened when his fantasy went on the air.

"How odd," he said. "I don't think I would care to comment until I hear more about it."

He wrote the "War of the Worlds" in 1898. It is often credited with inspiring most of the current craze for highly imaginative fiction.

The reaction of the public seemed almost as fantastic as the play itself to Orson Welles, the 23-year-old stage producer, who adapted the script from the Wells book and portrayed the principal character actor.

"We've been putting on all sorts of things from the most realistic situations to the wildest fantasy," Welles said, "but nobody ever bothered to get serious about them before. We just can't understand why this should have such an amazing reaction."

"It's too bad that so many people got excited, but after all, we kept reminding them, that is wasn't really true."

WABC explains

At 10 P.M., WABC sent out the following explanation of its "War of the Worlds" broadcast:

"For those listeners who tuned in to Orson Welles' Mercury Theatre on the Air broadcast from 8 to 9 P,M tonight, and did not realize that the program was merely a radio adaptation of H.G. Wells' famous novel, 'War of the Worlds,' we are repeating the fact, which was made clear four times on the program, that the entire content of the play was entirely fictitious."

Britain Had a Radio Panic Back in '26

It's happened before.

In 1926 Father Ronald Knox broadcast from Edinburgh through British Broadcasting Company a burlesque called "Broadcasting the Barricades," in which he simulated the part if an announcer of news bulletins and gave out hair-raising and preposterous news.

Although the nature of the program was stated beforehand, newspapers and the broadcasting office were flooded with telephone calls asking: It is true the House of Commons is blown up? What has happened to Big Ben?"

Some listeners fainted, and it required announcements by the broadcasting company and reassurances in the press finally to quiet the nation's fears.

http://www.nydailynews.com/entertainment/tv-movies/daily-news-original-coverage-war-worlds-article-1.305067

Name ______________________ **Course** ______________ **Date** ________________

Chapter Question

1. With the availability of modern technology, do you think that this type of situation could occur in the contemporary society? Explain.

TELEVISION, CABLE, AND THE FUTURE

Brent D. Ruben, Paul Reis, Barbara K. Iverson and Genelle Belmas

© wavebreakmedia/Shutterstock.com

Introduction

Since television entered people's homes in the 1940s, it has been a primary source of entertainment for many American families. Whether you watch one hour or 20 hours of television a week, the medium most likely plays some sort of role in your life. You may have learned your ABCs from a children's program, watched your favorite sports team win a championship, or witnessed the Twin Towers go down on September 11th through the power of the television screen. Imagine if you were unable to connect to the world in this way. How would you stay informed or entertained? While newspapers and radio can provide adequate information, they do not have the multi-sensory effect of television. You can listen to a space shuttle launch on the radio and hear the command tower count down to blast off. In a newspaper, you can see an amazing picture of thick smoke building up beneath the shuttle as it lifts into the sky. However, television has the power to bring sight and sound together to create an experience that makes viewers feel like they are witnessing an event in person.

Television often gets a bad rap. It has been given names like the "boob tube" or "idiot box", and those who are avid television watchers are often categorized as "couch potatoes". But television is much more than meaningless images with synchronized sound. It is a portal to a world that one would not see if not for the connective power of television. It is also an amazing piece of technology that required the efforts of many innovative individuals.

To understand how television has become such an important part of our culture, let's review the birth of the medium, examine how programming is developed, and discuss how new technologies have transformed the nature of television.

The Advent of Television

Television has taken many different shapes over the last century. Like many forms of media, it started as merely a concept and was developed into an industry. This industry was almost overlooked, as many individuals felt society had no use for television.

The Early Stages

While developing the technology to bring live pictures to a screen was a long and meticulous process, one of the first inventions that made television broadcasting possible came about accidentally. In 1872, English telegraph worker Joseph May made an unusual discovery about the element selenium. While examining materials for use in the transatlantic cable, the first telegraph cable to cross the Atlantic Ocean, May randomly placed a strand of selenium wire on a table near a window. As beams of sunlight passed through the window onto the wire, May noticed that the sunlight caused the electric conductivity of the selenium wire to change. Upon further investigation, he realized that the amount of electricity produced by the wire was in direct relation to the amount of light that fell on it. While this discovery was only of minor importance at the time, it provided the foundation for **photoelectricity**, or turning light into an electric signal.

Over the next few years, various inventors such as George R. Carey attempted to use this discovery to create electric images. They used selenium cells to break up a picture and sent the pieces over a wire to small electric lamps. Once the selenium cells reached the electric lamps the image would be recreated in a mosaic. A **mosaic** is a television tube that consists of many tiny photoelectric particles that convert light to an electric charge.[1] To recreate the image, the electric lamps would be turned on, turned off, intensified or dimmed. The more lamps used in the mosaic, the more detailed the picture. However, this process required an outrageous amount of wiring and an impractical setup of electrical lamps. To reach the level of detail in a modern day 23 inch television, it would take 350,000 lamps, each no more than one-fortieth of an inch in diameter.[2]

In 1884, German inventor Paul Nipkow avoided the problems of a mosaic device by applying the principle of scanning. **Scanning** involves viewing all the elements of a picture successively, rather than all at once, and sending them in a specific order over a single circuit. In this principle, a photoelectric cell would scan, or look over, only one portion of a picture at a time—going right to left then down a row, just as a person would read a book. Nipkow developed a rotating disk,

[1] mosaic. In *Merriam-Webster Online Dictionary* at http://www.merriam-webster.com/dictionary/mosaic. Accessed November 13, 2008.
[2] Christopher H. Sterling, John M. Kittross, *Stay Tuned: A history of American broadcasting*. Lawrence Erlbaum Associates: 2002.

called a scanner disk or Nipkow disk, which allowed information-filled cells to be sent from a transmitter to a reception point where the original image is recreated line by line. At the time, this technology was viewed as impractical. However, the concept of using a single wire to transmit an entire image became, and still remains to be, the basis of all television.[3]

More than forty years later, the concept of scanning was used to create the first successful televisions. While the machines may not be considered televisions by today's standards, they could transmit pictures. In 1922, Charles Francis Jenkins of the United States created a machine that sent still pictures through radio waves, but perhaps the more true form of television was created by John Baird in 1925. Baird's machine created a transmission of a live human face. The inventions of Jenkins and Baird were not immediately hailed as revolutionary technology. Critics felt the cost of building such systems would never be repaid and that the world did not have much use for television. Still, Jenkins and Baird, as well as companies like American Telephone and Telegraph Company (AT&T) and General Electric Company (GE) worked to further develop television technology. In 1928, Jenkins began to sell television kits that allowed consumers to build their own televisions. He also established his own television station and produced animated programs. In 1929, the British Broadcasting Corporation (BBC) allowed Baird to produce and broadcast a half-hour television show. These programs were made for extremely small audiences, as very few people owned television sets. However, in the following years, audiences would grow from tens to thousands as viewers were purchasing or constructing their own television sets just to catch a glimpse of a few flickering images on a dim receiver screen.[5]

Steady Progress

Determining who invented the television is a difficult task, as the efforts of many inventors played a role in the development of technologies used in a television system. However, two individuals played a major role in developing television into a mechanically sound and economically viable

Television

The word television was coined by a Frenchman named Constantin Perskyi at the 1900 Paris Exposition. On August 25, 1900, he read a paper to the International Electricity Congress and used the word to describe a device that could send images through electricity. Strangely enough, at the same exposition, a set of trading cards, titled "In the Year 2000" displayed images of inventions that were expected to emerge by the year 2000, one of which resembled a television. Little did they know that the television would reach their homes in only a few decades.[4]

---

[3] television. *Encyclopedia Britannica* at http://www.britannica.com/EBchecked/topic/1262241/television. Accessed November 13, 2008.

[4] Television History: The First 75 Years. (2006). *TV History*. Retrieved from http://www.tvhistory.tv/1890s%20Victorian%20Trade%20Card.htm.

[5] television. *Encyclopedia Britannica* at Encyclopædia Britannica Online: http://www.britannica.com/EBchecked/topic/1262241/television. Accessed November 13, 2008.

system. Vladimir Zworykin was a Russian immigrant who worked at Westinghouse Electric Company, a subsidiary of Radio Corporation of America (RCA) in Pittsburgh, Pennsylvania. In 1923, while working for the company, Zworykin filed a patient application for an all-electronic television system. Despite the fact that Zworykin was unable to build and demonstrate the system due to lack of funding, he managed to convince David Sarnoff, vice president of RCA, to provide $100,000 in funding on the promise that in two years he would produce a usable electronic television system. As Zworykin worked on building his system, another inventor by the name of Philo Taylor Farnsworth was demonstrating his basic version of an electronic television system called an Image Dissector. Farnsworth was an ambitious young man from Utah with only a high-school education. He learned about electronics though technology magazines and his own experimentation. Farnsworth secured funding for his project by gathering investors who hoped to gain a profit from marketing an affordable television system that was consumer friendly. While Farnsworth was able to develop his system for only $5,000, a fraction of Zworykin's cost, it would take many years and nearly $50 million dollars before anyone could create a television that was profitable.[6]

With the initial funds provided by RCA, Zworykin developed the Kinescope, a cathode-ray receiver that could project images onto a glass tube. **Cathode rays** are beams of electrons used to project images on to florescent screen inside a vacuum tube. Meanwhile, Farnsworth worked on perfecting his Image Dissector. A functional television could have been created more quickly and for less money if the two men were able to collaborate, but competition between RCA and Philco, Farnsworth's partnering company, kept the two innovators apart. In 1930, RCA attempted to purchase the patent to Farnsworth's Image Dissector for $100,000, but was turned down. RCA and Zworykin were able to learn enough from the Farnsworth's Image Dissector to develop the Iconoscope camera tube, which became their first working electronic television system.

This development created conflict between RCA and Farnsworth. In 1932, both parties met in court to determine who invented the electronic television. After a long process, the court eventually ruled in favor of Farnsworth, and in 1939 RCA agreed to pay royalties to Farnsworth Television and Radio, Inc, the firm Farnsworth created after leaving Philco. Still, RCA was able to gain a fair share of the credit for the television through its various publicity events. Later that year, RCA demonstrated its system at the 1939 World's Fair in New York City. At the official opening ceremonies, Franklin Delano Roosevelt became the first U.S. president to be televised. Some media outlets hailed television as a technology that would change the American home, but others, including *The New York Times*, described it as a gimmick without useful applications.

Television Reaches the Public

In 1941, the Federal Communications Commission (FCC) decided that the technology was well developed enough to be introduced to the public. In that year, fewer than 10,000 regular viewers were able to receive television transmissions from a handful of stations. Television sets were a very

[6] Ibid.

significant investment for a middle class family. A black-and-white receiver with a five-inch screen cost about $1,000, at a time when a Chevrolet cost $2,000. Still, the public was growing curious about television. Unfortunately, the attack on Pearl Harbor and World War II greatly delayed the widespread adoption of television by the public, partly because most electronic companies diverted their resources for the war effort. Many American television stations reduced their schedules or simply went off air. American electronic companies such as Bell labs, General Electric, and Westinghouse focused their efforts on using television to guide missiles, long-range reconnaissance, and other military endeavors. RCA's David Sarnoff proclaimed to President Franklin Roosevelt that, "all our facilities and personnel are ready and at your instant service."[7]

The war ended in 1945, and television expanded rapidly. By 1948, nearly 70 television stations were on air. The expansion was quickly halted by a second delay. Due to concerns regarding the limited space available in the VHF (very high frequency) band, the FCC administered a 4-year freeze on all new station licenses. In 1952, the organization opened the UHF (ultra high frequency) band to allow more space for television transmission and resumed issuing new licenses. By 1955, the cost of a television began to drop thanks to mass production and competition in the industry. That year, nearly 67 percent of Americans were equipped to receive programs in black-and-white. That figure grew to 87 percent by 1960.[8]

Television in Color

Just like black-and-white television, color television gained popularity slowly. In the 1940s, Peter Carl Goldmark of the Columbia Broadcasting System (CBS) introduced the first color television system. Subsequently, RCA developed a technique for color broadcasting that, unlike CBS's version, would be compatible with black-and-white broadcasting. This type of compatible broadcasting was approved by the FCC in 1953. Within the next year, RCA started to manufacture color television sets known as the CT-100 color receiver. The high price and questionable quality of color television sets caused the majority of consumers to stick to black-and-white sets. But over the next ten years, the price of color TV sets dropped and the quality improved dramatically. By 1967, color television sets outsold black-white-televisions for the first time, and in 1973 more than half of American households had a color set. Today, 98 percent of Americans have a color TV in their home.[9]

Television Goes Digital

In December 2005, the U.S. Congress approved new legislation requiring broadcasters to end all analog transmission of television signals. As of February 2009, any viewer using an antenna to receive television signals would need to acquire a converter box in order to receive programming.

---

[7] RCA Press Room, About RCA - Television, Retrieved at http://home.rca.com/en-US/PressReleaseDetail.html?Cat=RCAHistory &MN=9.

[8] http://www.history.com/encyclopedia.do?articleId=203824.

[9] Associated Press. "Color TV Hits 50th Anniversary in U.S." *USA Today*, March 24, 2004. Retrieved from http://www.usatoday.com/life/television/news/2004-03-24-color-tv_x.htm.

Digital TV transmission, already a reality for many cable and satellite service providers and consumers, promises better image and sound quality. The approved legislation also allocated up to U.S. $1.5 billion for consumer purchase of converter boxes that will allow all analog TV sets to receive the new digital signal.[10]

Digital television transmission, coupled with technological innovations on the receiving end, such as high definition television (HDTV), liquid crystal display (LCD) and plasma screens, and high quality audio systems, are poised to revolutionize the way most people watch TV. These new technologies have prompted a growing number of consumers to set up sophisticated home-theaters—entertainment centers comparable to the best theater-going movie experience.

The development of the technology behind television sets and television broadcasting took many years and numerous people to bring the technology to its present state of development. Similarly, the development of programming for television was, and still is, a painstaking process.

The Formation of Television Programming

While creating the technology that made television possible was a tremendous feat, deciding what should be broadcast on television was an equally challenging task. After the initial excitement of seeing live people appear on a screen wore off, it was apparent that quality programs would have to be developed if the medium was going to continue to grow and surpass radio. Different television genres and formats have been popular over the decades. While some of those faded into oblivion, some continue popular to this day, while others have disappeared just to reemerge later completely reformatted.

From Radio to Television

Early radio programs—variety shows, dramatic productions, comedies, and quiz shows—were performed live for simultaneous transmission throughout the country. So important was the supposed dynamic of a "real" event that announcers at one radio network were required to wear tuxedos; the management was convinced that this made them speak with distinction and moreover it both impressed the studio audience and sparked their earnest participation.

When many of the popular radio programs and formats moved to television, the tradition of presenting them live continued, if for no other reason than because it was quicker and cheaper than filming and editing them. Newscasters had to learn to look at the camera from time to time, but basically they were the same radio news readers with a microphone and a sheaf of papers in front of them, with little in the way of graphics or film footage to illustrate what they were reporting.

---

[10] Digital TV 2009.

Television as Entertainment

Variety shows were a natural for television. Whereas radio had been limited to presenting musical acts that could be appreciated by the ear alone, television could use baton twirlers, jugglers, tap-dancers, wire-walkers, puppeteers, and animal acts. **Variety shows** were presented live, sometimes from a hastily refurbished Broadway theater with an announcer standing in front of the curtain and ballyhooing the next act much as a vaudeville host might have done. Ed Sullivan's *Toast of the Town* long reigned as television's quintessential variety show. *Arthur Godfrey and His Friends* was an extended family of young performers chaperoned by a father-figure host, while Milton Berle's popularity stemmed from the comedian's willingness to go to any extreme to pull off a ridiculous skit. These types of programs paved the way for modern sketch comedy programs. **Sketch comedy shows** often contain a series of short comedy scenes, referred to as sketches. Sketches are about five to ten minutes long and typically performed by ensemble casts. NBC's *Saturday Night Live* is an iconic sketch comedy show that has aired for over 30 years. Variety shows have made a bit of a comeback in recent years with shows such as *American Idol* and *America's Got Talent*, but modern shows have added a competitive twist to the formula.

Dramatic television also made a smooth transition from radio to television. Radio drama had been fairly simple to produce live, because actors only needed to sound like the characters they played, and sound effects took care of setting the scene. Nonetheless, drama made the move to television with considerable success. A new generation of playwrights, directors, designers, and actors presented **anthology dramas**, high-quality original hour-long and 90-minute plays weekly on such shows as *General Electric Theater*, *Playhouse 90*, and *Studio One*. Eventually these live dramas were replaced with prerecorded dramatic programs.

Some lament the passing of live dramatic television. But a look at those old shows reveals that their production values were quite basic, and the look was flat in texture. Today we expect more polish and excitement from prerecorded programs.

Still, some entertainment programs such as talk shows maintain a live or partially live format. NBC pioneered the talk show format, wherein personalities provided a blend of entertainment, information, and service material. *Today* was the prototype for the early morning news-weather-feature-chatter wake-up format, and *The Tonight Show* paved the way for countless desk-and-couch late-night programs in which a genial host does an opening monologue, interviews a handful of celebrities, banters with the band and the audience, and bumbles through a topical comedy sketch.

Early television broadcasting was carried for only a few hours each day, mainly in the evening when whole families could watch. As programming was expanded into the daylight hours, children's programming was added. **Children's programming** consists of television shows that are appropriate and often educational for children under the age of 12. Sitting on bleachers or benches in tiny, cramped studios, an audience of a few dozen kids would be entertained by puppets, perpetually cheerful adults, a clown, and a few zany characters. Buffalo Bob was the host-sidekick to the freckle-faced *Howdy Doody*, in one of the earliest children's programs. It became the forerunners of popular children's programs such as *Sesame Street*, *Barney*, and *The Wiggles*.

The situation comedy is a staple of television programming. A **situation comedy**, or **sitcom**, is a program that typically consists of a group of characters placed in a familiar setting or situation that ultimately generates humorous activity. Traditional sitcoms are filmed in front of a live studio audience and allow viewers to hear the emotional reactions of audience members. *I Love Lucy*, a popular situation comedy of the 1950s, was the first television program to be filmed and then broadcast at a later time. The show pioneered the three-camera set-up, with one camera in the center used for the establishing shot of the Ricardos and their neighbors in the living room, and the other two focusing on the faces of Lucy and the person with whom she was speaking. On this program the bedroom—if pictured in any episode—showed two separate beds so as not to violate the moral sensitivities of 1950s viewers.

Today, notice how often the "situation" in a situation comedy revolves around a family or group of characters gathered around a familiar setting—the living room or a bar or an office or a coffee house—and the three-camera set-up moves the action along swiftly in a manner that is easy to follow on the small screen. This classic form of situation comedy is used in shows such as *Two and a Half Men* and *Everybody Loves Raymond*. Other shows such as *The Office* and *Scrubs* have scrapped the studio audience and added more cameras to create a program that is more like a movie than a television show.

Other popular programs that have evolved over the years have been detective shows and westerns. The shows are typically filmed on location and edited into a one-hour format, with commercial breaks nicely worked in at planned pauses in the action. Crime dramas joined sitcoms as the main programming for prime-time television. Networks and their advertisers wanted to train audiences to watch the same programs regularly, so they encouraged producers to develop familiar and likable characters who could be shown in standardized yet slightly different situations week after week.

Game shows were popular programs on radio, but they became even more popular on television as the agony or glee of the contestants could be seen in all its human glory. Easy and cheap to produce, the games became not only words, bells, and buzzers, but giant boards with flashing lights, handsome hosts, and maybe even beauty queens modeling or presenting the prizes.

For a short while game shows disappeared from television when it was discovered that some contestants on the most lucrative prime-time shows, including *The $64,000 Question*, had been coached in ways that prepared them to answer difficult or obscure questions. The quiz show scandals of the 1950s shook the public's confidence in the integrity of television programming. While it may have taken longer to win back the public's trust, games shows returned to television after four years. Today nationally syndicated shows such as *Jeopardy* and *Wheel of Fortune* are still extremely popular, attracting large audiences on a nightly basis. In the late 1990s, U.S. networks successfully imported and adapted European game shows such as *Who Wants to Be a*

Millionaire, which gained daily slots on ABC's prime-time schedule, and even topped the most-watched charts for a short time.

Another format that made the transition from radio to television was the soap opera. **Soap operas** are serial stories told through a series of characters, usually in installments. Named "soaps" because their earliest sponsors were often personal hygiene and house cleaning products and companies, these radio dramas thrilled generations of listeners before making their move to television. Initially, because of their storylines and daytime broadcast schedule, the characters and plots were of interest mainly to young and middle-aged women. But as storylines diversified and diverse characters were brought in, many men became big fans of television soaps as well.

The addition of visual components to the soap opera meant that several story lines could be followed at once. Along with the usual middle-aged characters developed in the radio format, television added plots involving more glamorous young people. Soap operas became so popular for a while that, in the 1980s, producers and networks moved to nighttime with dramas such as *Dallas* and *Dynasty*, which explored more mature themes. Many of the weekly one-hour prime-time dramas that were created in the 1990s such as *Melrose Place* and *Dawson's Creek* and up to today, such as *Desperate Housewives* and *Gossip Girl*, were heavily influenced by the original soap opera format and language.

Televised sporting events are one of the main features that set television apart from all other media. Sports events are complicated in terms of camera placements, following the action, and

Quiz Show Scandal of 1959

The most notorious case of game show rigging occurred in 1959 on the show Twenty-One. The program required contestants to go head-to-head in a trivia contest and answer enough questions to reach 21 points in order to receive a cash prize. Charles Van Doren, an English professor at Columbia University, was a celebrated winner on the popular game show, winning more than $130,000 and many admiring fans, but he soon found himself in the center of controversy. Around the time of Van Doren's win, rumors began to spread that game show executives fed certain players answers to questions in an effort to heighten drama and allow a more likeable contestant to win. During a congressional investigation, which was coordinated to examine the rumors, Van Doren admitted to cheating during his 15-week run on the show. The scandal followed Van Doren for many years, forcing him to retire from his position at Columbia and publish books under a pseudonym.[11]

[11]The American Experience, Quiz Show Scandal, People & Events, *Public Broadcasting System*. Retrieved from http://www.pbs.org/wgbh/amex/quizshow/peopleevents/pande02.html.

Telenovelas

Soap operas are popular not only in America but around the world. Latin American countries have been very successful developing their own brand of soaps, called *telenovelas*. These serial dramas, developed in countries such as Brazil, Mexico, and Venezuela, are exported all over the world, and shown on prime-time in countries as diverse as China, Turkey, or Italy.

cutting from close-ups to the entire field of action. But they were quickly embraced by television programmers because they attracted large audiences; they were fairly inexpensive to produce; and advertisers liked the tie-in with popular hometown teams. In addition to baseball, which radio had covered for decades, television elevated professional sports to new stature. It also brought wide popularity to boxing, professional wrestling, bowling, and even marginal sports-entertainment attractions such as the roller-derby competition.

Reality programming is one of the newer forms of television programming and has become a very successful genre. The term **reality TV** is generally applied to describe programs that focus on non-actors that produce unscripted dialog. The format has been so victorious on TV that it has gained its own Emmy category. Although many question the label itself, since it has revealed several circumstances in which reality was tweaked to produce more desirable results, reality shows have made a strong mark on the network's schedules, and at least at this point don't seem to be losing steam.

Ranging from the prosaic—talent shows such as *American Idol*—to the bizarre—survival shows such as *Survivor* and *The Amazing Race* that test contestants' mental and physical strength. These reality shows are relatively cheap to produce and generate enough buzz to gain their own free publicity. Many analysts have also credited their success to the fact that they give audiences the sense of being able to watch themselves through the eyes and experiences of "real people."

Television as an Informer

In the early stages of development, many supporters viewed television as a revolutionary way to inform and educate the masses. Today we take for granted the ability of network and local news departments to provide complete (and many times live) coverage of presidential elections, natural disasters such as the 2004 tsunami, space shuttle launches, terrorist attacks, sport events, wars, inaugurations, and just about any other newsworthy event. Yet, these news functions were developed and perfected in little more than one generation.

Early television was not capable of providing much more than the 15-minute reading of the news that radio offered. Gradually, film footage and then live transmissions were added. Because of the addition of visual images, news broadcasts were allotted more time. By the end of the 1960s, first with the civil rights movement and then with the Vietnam War, television news had guaranteed a significant, permanent, premium (and much longer) space in the networks' schedules.

Coverage of the political nominating conventions marked the 1950s and made the careers of seasoned journalists such as Edward R. Murrow, David Brinkley, Mike Wallace, and Walter Cronkite, who had all migrated to TV from traditional media such as radio and newspapers. The presence of TV cameras in those conventions made them not only more relevant for the audience, but also made the medium much more attractive as a workplace for those seasoned reporters. Television made the nominating conventions a prime-time show, just like the World Series of baseball or the Miss America Pageant.

Because of their prominence at historic events, the network anchors were elevated to positions of power and respect unusual for anyone in American society. That's why Walter Cronkite, who consistently had the highest believability ratings before his retirement from the anchor seat in 1981, came to be known as "Uncle Walter" for the American public. The viewers felt he not only brought them the news, he in fact embodied the news.

In one decade or so, the era of the news reader was over. The all-powerful anchor and the participant journalist were now the norm on TV. The same way cameras, and later transmission dishes, made images and live transmissions an essential part of television journalism, it quickly became apparent that the anchors and other on-air personalities didn't have to remain tied to the New York or Washington studio. Instead they could be expected to travel with the President to a summit, attend the Olympic Games, and even visit the war front as Cronkite and others memorably did during the Vietnam War.

Some news programs began to take an unlikely turn in the 1980s, as infotainment became a hot new format. **Infotainment** is a form of programming that combines current news and feature stories. Programs such as *Entertainment Tonight*, and *Extra*, presented soft celebrity or crime news. The fast-moving, easily digestible format blended the structure of a news program with the easygoing style of a talk show. The anchorpeople are encouraged to be personalities themselves, and to feature the same stars that the audience read about in *People* and other similar news-gossip magazines.

It was just a matter of time before the success of such infotainment shows rubbed off into the traditional news shows. Currently, even the networks' serious news shows, such as *60 Minutes* or *Primetime Live*, have either generated their own spin-offs or incorporated many of the blended features popularized by infotainment. As recently as ten years ago, no one could imagine that the traditional news magazines would be racing to scoop each other on topics such as celebrity infidelity, Britney Spears, or New York socialites. Many in the news business defend themselves by pointing toward the audience, and saying that they are only giving the public what it wants.

Walter Cronkite and Tim Russert

Nothing attests as much to the power of the television newsman or newswoman than our ability to recall important historical events based on how well-known journalists delivered those events into our living rooms. Two recently deceased television journalists, Walter Cronkite and Tim Russert, have come to symbolize (or even become synonymous with) television news for several generations of viewers.

When Cronkite died on July 17, 2009, the *New York Times* noted in its obituary how "from 1962 to 1981, Mr. Cronkite was a nightly presence in American homes and always a reassuring one, guiding viewers through national triumphs and tragedies alike, from moon-walks to war, in an era when network news was central to many people's lives." For many of our parents and grandparents, Cronkite's self-assured and reassuring demeanor became indelibly associated with some of the important events of their generations. His passing at age 92 reminded many of the golden age of television news, when nightly newscasts neatly summarized the news of the day, and no one could have anticipated an era of 24-hour news cycles and countless news blogs.

Tim Russert's death a year earlier at the age of 58 marked a different kind of passing. For many television observers, if Cronkite had a successor, his name was Russert. With his sharp and inquisitive mind hidden behind an affable Midwestern persona, Russert was able to deliver (and discuss) political news with the same kind of gravitas Cronkite employed to interview presidents and announce life-changing events. In its obituary of Russert, the *New York Times* called attention to the fact that he had come to play "an increasingly outsize role in the news media's coverage of politics." It may well be the case that Russert's death will come to symbolize the end of "personalized" political coverage. As blogs and Web sites multiply endlessly, it becomes harder to imagine any one political journalist coming to embody the "everyman quality" Russert brought into television news.

Similarly, many news talk shows and news segments have become glorified soap boxes for celebrities to promote their latest films, books, or CDs. Shows such as the highly successful *Today* and *Good Morning America*, which are currently competing for the top morning ratings spot, have also incorporated infotainment into their format, transitioning easily from serious news segments to celebrity talk to cooking and fitness tips.

Since the events of September 2001, as the general public turned once again to the networks for reliable information, broadcast TV's news audience seems to have stabilized, after a brief increase in numbers. The competition from 24-hour news channels such as CNN and Fox News spurred the networks to reinvest in their coverage, a fact that became clear during the first stages of the Iraq War, when every single major news outlet had their own ground crews—and sometimes their anchors and star reporters—in the battlefield.

Television Networks and Public Broadcasting

Each of the major radio networks entered the television business in the late 1940s, with NBC and CBS signing up affiliates in most of the 100 major markets, while ABC and Dumont—the long-forgotten fourth network—lagged far behind. Dumont dropped out of competition in the mid-1950s, and ABC became the dark horse in the network field, a role that it would hold until its ascendancy in the 1970s as the premier sports network.

Early Network Programming

In the early stages of the medium, there was insufficient programming for round-the-clock broadcasting. Typically, the station of the late 1940s and early 1950s came on the air with a "test pattern" around 5 p.m., carried one or two local programs, including the reading of the news in the early evening, and then joined the affiliated network for three hours of programming—the beginning of the concept of prime time. That paralleled radio network programming, and the format remained the heart of the network-affiliate concept even after the networks began to provide soap operas and game shows during the daytime hours.

Because the art of television production would take time to develop, each network carried movies that filled almost the entire three-hour prime-time slot at least one or two evenings a week. Each network began to work with advertisers and film producers to develop continuing half-hour and one-hour programs. Soon certain nights took on a distinctive flavor. Saturday emerged as comedy/variety night, with CBS's popular *The Jackie Gleason Show* and NBC's *Your Show of Shows*. CBS established several situation comedies on Tuesday night and the other networks accepted its domination in that segment of programming, much the way in which the other networks conceded in the 1990s and early-2000s that Thursday nights were NBC's "Must-See TV" nights, where the network reigned unchallenged with its popular sitcoms.

The networks gradually developed the capability of filling every night of the week with original programming, and they had to do it 52 weeks of the year because they had no ability to store and reuse television programs. Filmed "kine-scopes" of television programs were useful for archival purposes, but they were not of broadcast quality. The introduction of the videotape—an electronic means of recording and storing television programs—meant that the networks could make just 26 episodes of a weekly program and show "reruns" during summer months.

By the 1960s, the networks had developed the practice of introducing most of their new shows one or two weeks in the middle of September. Shows that did not find an audience fairly rapidly were canceled and replaced with other programs whose pilot episodes shown the summer before did not win them a slot in the opening fall line-up. This practice is still used by the networks, with the possible difference that now unsuccessful shows typically get canceled much faster than they did before.

© 1000 Words/Shutterstock.com

Late night talk shows have been a successful and enduring staple of network television. The most famous and longest running of them, the *Tonight Show* on NBC, started in 1954, when the network decided to broadcast nationally a local New York City talk show with host Steve Allen (*Television Talk: A History of the TV Talk Show*, by Bernard M. Timberg and Robert J. Erler, University of Texas, 2002). The genre was a direct successor to similarly formatted radio shows. Entertaining and cheap to produce, late night talk shows proved to be a great success with audiences, mixing sketch comedy, stand up bits, live music and celebrity interviews conducted by an affable, easy-to-relate-to host who normally stayed away from controversial issues and hardball questions.

Hosted successively by Steve Allen, Jack Paar, Johnny Carson, Jay Leno, Conan O'Brien, and Jay Leno again, the *Tonight Show* generated other successful imitators in competing networks, *Late Night with David Letterman* being the most famous of them. Written off by many as a dying format, the late night talk show genre has, time and again, showed its enduring appeal and relevance—throughout the 1990s with the rivalry between Jay Leno and David Letterman, and more recently with the addition of O'Brien into the mix. The format has also sparked a sub-genre in the political satire talk show category, with Comedy Central's *Daily News with Jon Stewart*, and the *Colbert Report* with Stephen Colbert, and HBO's *Real Time with Bill Maher* creating a whole new generation of viewers and admirers.

FCC Examines the Role of Networks

Ever-mindful of its mandate to prevent monopoly control of the information carried on the public airwaves, the Federal Communications Commission began the first of several inquiries into program production and distribution in the late 1950s. In 1970 the FCC passed the "Financial Interest and Syndication Rules," a document that forced the networks out of the business of syndication. A network could no longer control the profits from reruns of its shows. Shortly thereafter, the federal government filed antitrust suits aimed at curtailing network control of programming. By the end of the 1970s, all three networks had signed consent decrees that limited the number of programs they could own. Syndication and program control were profitable for the networks, so the FCC limitations, while they prevented a monopoly in the production and distribution business, also precipitated the decline of network revenues.

Indeed, by the beginning of the 1990s, as the share of audience and the revenues from the networks plummeted, the FCC began to relax its restrictions on the networks. The Fox Broadcasting

Network Affiliates

Companies such as NBC, owned by RCA, transferred an adapted version of the radio network system studied in the previous chapter to the newly-developed medium of television. The very profitable and accommodating corporate parent-affiliate relationship proved to be an even greater success on TV. In the largest and most profitable markets, such as New York and Los Angeles, the corporate parent also owned the station, but ownership limitations imposed by the FCC left the ownership of most stations to local, and at the time family-owned, companies.

The affiliation system has survived for so long because it is based on a close-knit relationship beneficial for both parent network and affiliates. Through this system, local stations receive network programming from the parent station (which they rebroadcast for a fee); produce local programs such as regional newscasts; and also purchase nationally syndicated programs, which help to fill the stations' daily schedule. In return, besides paying fees for the shows they receive, affiliates also pay to the parent network a percentage of the income generated by local advertising, and receive a percentage of the nationally broadcast commercials. In many cases, particularly in special programs and sporting events such as the NFL and NBA seasons and the Olympic Games, network and affiliates strike deals on how to finance the costs and divide the profits. These deals are not always easy to arrange, but the survival of the affiliation system speaks for its mutually beneficial aspects.

Company, bidding to become the fourth network, was granted a waiver by the FCC to program 18 1/2 hours a week without being subject to the syndication rule.

In 1991, the FCC made some concessions to the networks, allowing them to keep resale rights to as much as 40 percent of the programming they produced, and to earn royalties on reruns of other programming. From 1991 on, the FCC has reviewed many of its ownership rules in a way that greatly benefited the networks and the parent companies that own them.

Networks Loss Profits

The networks began the 1980s with a combined share of 91 percent of the television viewing audience. By the early 1990s they had about 60 percent. By the turn of the century, the networks had lost 44 percent of the 1980 nightly TV audience to other media such as basic and premium cable as well as satellite, dropping their combined share to approximately 50 percent of the audience.[12] This decline translates into lost advertising revenues, because shrinking audiences put the advertisers in a much stronger bargaining position when they haggle with networks over

[12] "What's Really Going On In The Media World" by E.W. Brody. *Public Relations Quarterly*, Vol. 49, Issue 1, pg. 9, 2004.

"upfront" sponsorship of shows—committing themselves in advance to pay for advertising that will put a show on the air. This also signaled another significant phase in the demassification of a unified mass audience for TV.

The networks have struggled hard not only to stop this audience decline but to regain some of the lost market share. They have also tried to find different ways of increasing their profitability. Inventing or "reinventing" genres and formats such as reality TV and game shows has been one of these tactics. By creating or betting on these formats that are cheaper to produce and seem to attract new viewers, networks hope to stop some of their losses.

Another area that has seen a lot of investment—and a great deal of return for television channels—is DVDs and digital media files. These pieces of technology are a cash cow for networks and cable channels, as well as the programs' producers. They are very cheap to produce and release—recent and even "classic" sitcoms, dramas, miniseries, and TV movies have already been produced and only have to be repackaged. In some cases, programs are loaded with extras and bonuses, such as cast commentaries and deleted scenes. The costs involved usually relate additional copyrights and bonuses for the cast, creators and producers, as well as marketing and publicity.

Public Broadcasting

Not all forms of television broadcasting seek to make a profit. **Public broadcasting**, unlike commercial broadcasting, is meant to be used as resource for the public, not a means to turn a profit for a corporation. The advent of television was hailed as an opportunity for new and unparalleled efforts to educate the masses with the new technology. Unfortunately, commercial interests were so strongly opposed to reserving part of the broadcast spectrum for educational television that educational forces had to pressure the FCC in a systematic and sustained way. Once the network licensing freeze was over in 1952, the FCC agreed to reserve one channel out of every eight for noncommercial use.

While noncommercial programming does not generate revenue, it still costs money to produce. Therefore, funding is an essential part of its survival. Universities sponsored the first educational television stations, but financial support was difficult to find. The Ford Foundation, which helped educators put pressure on the FCC, continued its support by underwriting National Educational Television (NET), a consortium of nonprofit stations that formed their own sort of network. Rather than feeding programs electronically from a central location, as the commercial networks did, they merely mailed or delivered recorded programs to one another to rebroadcast.

By the late 1960s, there were nearly 200 educational television stations, and it was apparent that they needed to be put on sounder footing. The Public Broadcasting Act, passed by Congress in 1967, created the Corporation for Public Broadcasting (CPB) to foster the growth of nonprofit radio and television stations and facilitate their linkage for the purpose of sharing production costs. Two years later, the Public Broadcasting Service (PBS) was created specifically to facilitate nonprofit television programming. The CPB is a primary supporter of PBS.

The Public Telecommunications Financing Act of 1978 requires the CPB to submit to the president and Congress a "comprehensive and detailed inventory of funds distributed by federal agencies to public telecommunications agencies." Up until the 1980s, the federal agencies contributed the most to CPB's budget. Then, federal support for public television was seriously cut back. More and more, public television turned to funds from viewers and the business community. Pressured by a lagging income, in 1988 PBS changed its rules to permit the mention of one specific product or service each time a corporation received credit for funding a program.

Today, only about 14 percent of the funds that support public broadcasting come from the federal government. The rest is spread among state governments, the business community, and the general public, through personal donations.

Pressure on the Networks

The chief programmer at each of the major broadcast networks spends a lot of time studying the scheduling board on the office wall that shows the three hours of prime-time programming—8 to 11 p.m. in the East and West, 7 to 10 p.m. elsewhere—for seven nights a week—22 hours (Sunday nights carry an additional hour of prime time) for each network to fill, 88 hours of programs for the four major networks combined. Because about half will be 30-minute episodes of situation comedies and dramas, and each network offers at least one 2-hour movie each week, the total number of prime-time programs each network offers at any one time is around 30.

Trying to decide what will constitute an effective lineup of shows each night in terms of holding an audience from one show to the next is complicated enough. The problem is compounded by the necessity of competing with other major networks, independent stations, public broadcasting, cable, DVDs, Internet, video games, and other distractions. But *scheduling* is only part of the complex business of network programming. There are other pressures on the chief programmer, all of them pushing the network toward decisions that result in "lowest common denominator" programming that makes it difficult for anything but mediocre programs to survive the process.

Among the groups that affect the networks' decisions are:

- Advertisers
- The viewers
- Government regulators
- Advocacy groups and television "watchdogs"
- The competition—all television, not just other networks
- Packagers—studios and independent producers of programs
- Artists and creative people, many of whom control their own shows
- Affiliated stations

Any of these groups, when and if they become actively involved, have a veto power over the network. When you are dealing with such a large audience and such high production costs, you cannot afford to ignore or offend any segment of society. That's a dilemma that confronts network television more than any other medium.

Advertisers want popular programs that attract everybody and offend nobody. The stations that affiliate with the network have much the same outlook. In addition, the affiliates often have concerns that are regional: Stations in the conservative, religious regions of the country are more likely to drop a network show with too much emphasis on sex and violence or controversial topics, while outlets in the major urban centers might complain about programs that do not provide enough thought-provoking material.

Network Ratings

Every mass medium has a method for measuring the size and composition of its audiences. Audited circulation figures enable advertisers to know how many people read a newspaper or magazine, and surveys provide specific information about which people make up the audience. Box office figures measure theater attendance, and theater producers can observe the house to know which segments of the audience are responding to the program.

But broadcast media, as we saw previously, pose a special problem because the consumers can change the station they are watching from minute to minute, and there is no way for the broadcaster to observe or calculate which family members are in the audience for particular programs. A special form of research, then, is needed to measure who is watching what. In order for the information to be useful, it has to be quantified quickly for programmers and the advertisers they must satisfy.

What Is a Rating?

The firms that measure the reach and penetration of television programming know how many homes in the country have television. It is against this total number of potential viewing units that the ratings are calculated. The ratings for a given half-hour never add up to 100 percent, since 100 percent of homes with television aren't viewing at any one time. The top program in prime time may have a 30 rating, and all of its competitors in that time slot may have a combined 40 rating, for a total of 70—meaning that 30 percent of the sets in the area measured were not turned to any program.

Ratings are most significant in evening prime time, when most people are likely to be at home and free to watch television. This is the period when the major manufacturers of consumer goods and services want to reach the largest possible audience. Ratings are far less significant in the middle of a weekday morning, when as few as 5 percent of households use television.

At times of the day other than prime time, the measurement known as the **share** is more significant. It is possible for a local show to have a rating of only 7, which means that 93 percent of the area homes with television were not tuned in. But the show could have a 40 share, meaning that 4 out of every 10 sets that *were* turned on at that time were tuned to the program.

National advertisers of cars, soaps, and soft drinks that want to reach every consumer probably will not be impressed with a 40 share when they discover that it is coupled with a 7 rating. But a station that has the most popular program in the cheaper time slot will attract plenty of local and specialized advertisers.

Development of the Ratings Systems

The first ratings systems, developed for radio use in the 1930s, involved telephone interviews ("What program are you listening to at this time?), face-to-face interviews at the door ("What stations have you listened to in the past 24 hours?"), and even postcard reply surveys ("Please check off which stations were listened to this past week, and by which family members"). All of the systems relied on the cooperation of those who responded, and on their recall ability. Clearly a system that monitored broadcast consumption as it occurred would be preferable.

The AC Nielsen Company acquired the rights to a device, invented by two professors at MIT, called the **audimeter**, which could record when a radio set was turned on and to which station. When a family agreed to be one of the 1,000 Nielsen families across the country, they were visited monthly by a technician who removed the punched tape carrying the listening information. The data might be nearly two months old by the time Nielsen could assemble them for use by broadcasters and advertisers.

By the time television was introduced for home use, however, the audimeter had been improved so that the consumer could remove the tape and mail it in. Nielsen shifted from radio measurement to rating the television programs, and by the 1960s the technology had improved to the point where data was automatically fed to the company over phone lines, making the data available to the producers within a few days. Today, both Nielsen and its main competitor, Arbitron, supplement electronic data with the phone surveys in major markets when instant information is demanded by the industry.

Critics of the Nielsen technique point out that a television set may be put in use merely to provide background sound, to keep a pet or baby company, or to convince burglars that someone is home—with nobody really *watching* television; and thus no consumer paying attention to the advertising messages. The rival Arbitron system for a while required family members to make notations in a diary concerning the times when the set is on; to which channel, and, most importantly, which members of the family were watching. Critics questioned whether the diaries were kept carefully and accurately.

Neither method measured "why" or "how intently" the program was watched, or what value the audience put on the message after they had watched it. That is why the ratings firms augment their usual measurements with telephone surveys that go into greater detail.

Cable Television

The concept of *broadcasting* is exciting because information is beamed out for anyone to receive—a true and democratic *mass* communication system. Still, there are problems with broadcasting. The signal goes only so far, depending on the power of the transmitter. Signals can be lost if mountains or tall buildings intervene. Perhaps more importantly, the number of signals that can be carried on local airwaves is limited. Within 30 years of the advent of broadcasting, it was clear that technical improvements were needed in order to ensure that clear and adequate signals could be supplied to every consumer, and that many more choices could be added to the average consumer's television "menu."

Transmissions take barely a fifth of a second, and thus **superstations**—broadcast stations that decided to serve the entire country instead of just their local broadcast area by offering themselves to cable operators—in Atlanta, Chicago, and New York can be viewed in places thousands of miles away. Entertainment and news services also are fed to foreign countries, so the same movie, sports event, or 24-hour news can be viewed simultaneously in Detroit, Madrid, Cancun, or Hong Kong.

How Cable Works

At its simplest, an old-fashioned cable system consisted of:

- An *uplink*—the earth station that sends signals carrying programming up to a satellite. The satellite may then beam the information to yet another satellite serving another sector of the globe.
- A *headend*—the electronic center where signals are captured by an antenna. When satellite transmissions are involved, the headend is a *downlink*. Signals at the headend may be scrambled to prevent unauthorized use by anyone who is not a subscriber.
- *Trunk lines* that carry the signals into districts—a neighborhood, community, or small town—where feeder cables go down each street either on poles or underground, generally following telephone lines.
- A *drop line* to the individual customer. Inside the subscriber's house, a converter attached to the television set transforms the electronic signal into an unscrambled picture.

Cable Begins as CATV

By 1950, **community antenna television (CATV)** was bringing signals to remote areas and those previously cut off from broadcast centers by intervening terrain, such as the mountains of Colorado. CATV entrepreneurs erected tall antennas to capture distant or blocked signals, and then ran coaxial cable to the homes of subscribers who were happy to pay for the service. Eventually CATV operators began to "import" programs from beyond the area served by their antenna, and the concept of cable television began to develop.

The first technical problem occurred when the fledgling cable companies went to string their wires. Some planned to run the cables from house to house, but that was feasible only in areas of dense settlement. The obvious solution was to use the poles owned by the telephone company, as other utilities have done. But telephone companies set rates in many areas that added greatly to the cost of cable. Managers of the AT&T system forecast correctly that cable—far from being just a means of improving the television signal—was an alternative means of bringing many kinds of information into homes electronically. That put the cable companies in potentially competitive conflict with the phone companies.

Cable Growth

The flurry of interest in cable distribution of programming in the 1950s led some to predict that the entire nation would soon be wired. However, 50 percent penetration of American homes was not realized until 1988. One hundred percent cable availability was not a reality until the 1990s, when alternative systems such as satellite TV began to cover areas that were previously bypassed by traditional cable. Just as it did with broadcast television, the FCC has had to work its way through several difficult issues to ensure that all customers are served adequately and fairly, and to guard against domination of the field by any one economic group or interest.

In 1959 a U.S. Senate subcommittee considered legislation to require licensing of CATV operators, who were not regulated by any agency. Lacking information about the scope of cable and unsure of the many ramifications of the new technology, the legislators argued the bill furiously. At that point, states and municipalities jumped into the regulation of cable, forcing owners to obtain local franchises in order to string wires through or under the streets and requiring them to return a percentage of the income to the municipality in return for that franchise. The borough of Manhattan was among the first areas to be served by cable because its tall buildings make broadcast reception difficult. Many areas of New York City took longer to adopt cable because of the high costs. This situation was symbolic of the chaos and uncertainty of cable distribution.

Regulating Cable

The FCC did not get involved in regulating cable until 1965, when it ruled that programs imported from more than 60 miles away could not duplicate any local broadcast in the preceding or following 15 days. That restriction prohibited a cable operator, for example, from using a satellite to pick up a concert that was broadcast live in another market but that was scheduled to be shown on tape locally at a later date. This "syndicated exclusivity" regulation pleased the broadcasters but angered the cable operators, who joined forces to lobby against it. They succeeded in convincing the FCC to shorten the exclusivity period to just one day. It was the first event in an ongoing struggle that saw the cable operators gaining strength and forcing the FCC not to automatically favor the entrenched broadcasters.

In 1968, the U.S. Supreme Court rejected the argument by cable operators that cable was not interstate and therefore not subject to regulation by the federal government. Moreover, it said FCC regulation of cable was necessary in order to carry out the mandate of congress to develop the dissemination of electronic information in the best interests of the public. That decision laid the foundation for the FCC's heavy involvement in regulating cable.

In the 1970s, cable companies began to offer original programming, and they broadened the information they could offer by receiving signals from satellites. Now homes wired to cable could receive programs from anywhere. When cable differentiated itself as the "multi-channel medium" with the most variety to offer, it became attractive even to those in populated areas who did not need CATV in order to receive a signal.

The 1972 Regulations

The FCC issued a set of regulations in 1972 that put cable's house in order, superseding earlier decisions as well as a jumble of state and municipal regulations. Among the most important provisions were the following:

- *Certificate of Compliance*. While municipalities could still franchise cable owners, each system operator was required to obtain a "certificate of compliance" from the FCC. The granting of the certificate was virtually automatic, but the process alerted the FCC to verify that the operator was capable of providing service and that the municipality stayed within guidelines in setting reasonable franchise fees.
- *"Must Carry" Rule*. To ensure that cable operators would not carry just the most popular and profitable programs, all local signals had to be carried on cable if it is requested by a local station. The **must-carry rule** was popular with stations but unpopular with cable operators who preferred to pick and choose among local stations to maximize their profitability.
- *Importation of Signals*. Cable operators were permitted to import signals in order to carry all three major networks and adequate independent stations. There were several rules governing how many stations could be imported, and operators could not bring in a network affiliate hundreds of miles away that was available closer to the home region.
- *Anti-siphoning*. Cable systems were allowed to offer pay channels carrying movies, sports, and concerts. However, siphoning of events and programs that previously appeared free on broadcast television was prohibited. The FCC administered a rule in 1968 to prevent popular broadcast programs such as the Super Bowl, the World Series, and the Olympic Games from being expropriated by pay television.
- *Multiple Ownership*. Owners of television stations could not operate a cable franchise in the same community, and the television networks were prohibited from owning cable channels.
- *Local Origination*. Cable operators were required to provide local origination programming. In the 100 largest markets, these included "access" channels available to individuals or groups who wanted to get their information on the air.

The 1972 regulations required cable operators to provided original material "to a significant extent," but start-up costs made it difficult to build and equip studios. In the large markets where local access was required, franchisers provided little more than a rented storefront equipped with one or two cameras and tape recorders supervised by a single technician. Long-winded speeches by minor politicians, artistic presentations by those with marginal talent, and low-budget news programs have been carried by the public access channels in some large cities. On the positive side, schools and colleges have been encouraged to create and broadcast programs using their own television production facilities and crews.

Deregulation Gives Cable Greater Freedom

The 1984 Cable Communications Act eased regulation on the industry. It came after lobbyists for cable owners successfully convinced congress to shield them from controls by municipalities and states. The owners argued that the prime responsibility of the operator should be to the subscribers of cable television, not to government bodies, and that the market would determine the prices people wanted to pay for various services.

In 1985, the must-carry rule was struck down by a federal court and the FCC did not fight the move. The chair of the FCC said the decision "takes the first step toward a true marketplace for the distribution by cable systems" and "represents a positive step toward full First Amendment protection for all forms of electronic media." The demise of the must-carry rule occurred after Supreme Court Chief Justice Warren Burger refused to heed the call of broadcasters, especially owners of the smaller UHF stations, who wanted to ensure that cable had to carry their programming. The president of the National Association of Broadcasters called the action, "a blow to the localized system of broadcasting in this nation." Even some FCC staff members feared that greedy cable operators would ignore local needs and offer only the most profitable programming. New York City cable operators immediately dropped the New Jersey public broadcasting channel, although many New Yorkers work or travel in New Jersey and could be interested in news and programs from their neighboring state.

Cable operators proclaimed they would continue to fight regulation by the FCC and other authorities on the grounds that they should have all the freedoms accorded to the print media. They argued that they must have the flexibility to find programming that is different from broadcast fare. They also wanted to explore the concept of **narrowcasting**—offering channels for specialized audiences such as children, rock music fans, sports enthusiasts, and hobbyists.

In 1988, the Reagan Administration released a report by the Commerce Department's National Telecommunications and Information Administration that called for even further deregulation of cable in order to foster more competition. The report was aimed at ensuring that a few large companies would not dominate the industry. While it did not suggest that telephone companies should be allowed to enter the cable industry, it did suggest that telephone lines could be used as common carriers to help more cable companies gain access to information.

Re-regulation or Continued Deregulation?

In 1990, the head of New Jersey's Board of Public Utilities claimed that cable television in the state of New Jersey was a monopoly and suggested that permitting more than one cable company to operate in a town would benefit subscribers by giving them a choice of providers and lower rates. Cable companies argued that federal deregulation precluded states and communities from interfering with their franchises. While cable rates and programming policies cannot be dictated by states or communities, local authorities can set service requirements for utilities. New Jersey has strengthened rules that require quicker response to service outages and billing credit when service outages occur.

Bills working their way slowly through Congress at the beginning of the 1990s would allow cities to regulate rates in areas without two or more cable operators, limit the number of subscribers nationally that any one cable operator could serve, regulate the prices of premium channels, and reinstate the must-carry rule. Cable operators, who argued that they spent extraordinary amounts of money to develop almost a million miles of system in just one decade, nevertheless prepared themselves for some form of re-regulation. In the summer of 1991 the FCC reinstated local rate regulations for many cable operations. Working in favor of the cable companies was the fear on the part of many in congress that allowing states and municipalities to control rates and decide on the number of franchises would lead back to the corruption and chaos that marked the early days of cable regulation.

In the fall of 1992, the pendulum continued its swing back toward regulation of cable when both houses of Congress passed a bill to regulate rates for basic cable services and spur competition. Consumer groups and broadcasters were among the biggest supporters of the measure, which passed with sufficient majorities in both houses to override the expected veto from President Bush. Nevertheless, in January 1992, the Senate handed cable operators a setback by passing legislation that would strengthen both local and federal regulation to limit the increases in cable rates. Siding with broadcasters, the legislators brought back the must-carry rule in the Cable Act of 1992. The rule was upheld as both content neutral and necessary by the Supreme Court in two related cases in 1994 and 1997. Stations could either choose the must-carry option or a retransmission consent option. If they chose the must-carry option, cable systems had to carry those stations but did not need to pay for them. If the stations chose retransmission, they could be carried for some kind of compensation. Most large stations chose the retransmission option. The must-carry rule is being challenged in court again in 2010 by Cablevision, an east-coast cable company.

Cable Matures as a Business

Through the 1980s, the popularity of cable grew at a tremendous rate. More than 4,000 new systems were launched in that decade—a new system virtually every day. Initially many were owned by local entrepreneurs. Some had media experience, others were businesspeople eager to make a profit, and still others were engineers who had expertise in the technology. Eventually many of those that didn't have an understanding both of the technology and the business sold out to companies that did. Many of the new owners were parts of chains, and some of the largest chains were owned by large communication companies.

Westinghouse, the electric manufacturing company, for example, entered the cable business in 1981 by purchasing Teleprompter, the firm that wired the northern half of Manhattan in New York City. Just five years later Westinghouse put Teleprompter's parent company, Group W Cable Inc., up for sale and turned a profit. Westinghouse management had decided to concentrate on broadcasting, where the investment offers a quicker return to stockholders. Warner Communications (now part of Time Warner) formed Warner Amex, but a few years later bought out American Express's share in order to form a new cable venture with Viacom. Ownership patterns were in a state of flux throughout the 1980s as communication companies tried to figure out whether cable belonged in their mix.

Eventually chain ownership became the norm. The government's encouragement of further deregulation at the end of the 1980s was based partly on a concern that a few companies were cornering a large segment of the market. The largest, Telecommunications Inc., had 20 percent of the nation's 42 million cable customers. Deregulation was seen as encouraging more players to enter the market.

Cable Becomes Profitable

Among the largest cable services, only one was profitable in 1983: TBS, the Atlanta-based Turner superstation. Five years later, however, most of the top advertiser-supported services made money, led by TBS, Turner's Cable News Network (CNN), and ESPN. In fact, each of those top three cable services made more profit than CBS and ABC combined. Under deregulation, the networks were free to enter into ventures with cable services, and so the combination of players was constantly shifting throughout the 1990s and into the new century.

Cable's profitability can be exemplified by the increased costs of cable subscriptions. At one time the average consumer paid less than $10 a month for cable. Now, many subscribers pay a basic rate over twice or three times that much and add premium services that may bring the monthly bill to over $100, when taxes, fees and Internet services are added.

Eventually—assuming that the FCC decides to require competition in the market—cable will approach 100 percent penetration of the American market and consumers will draw the line on how much they are willing to pay and how many services they need. When that happens, the cable industry will "shake out," meaning that many companies will exit because they cannot make a profit, and the remaining companies will solidify their holds over the segment of the market they control. Rather than being producer-driven the industry will be equally consumer-driven. There will be a balance that involves information being created, priced, and delivered by the producers according to the needs and interests of the consumers.

Users Shape the Medium

Most mass media have to poll their audiences to find out what they like and don't like. Cable gets constant indications of the popularity of its services. Subscribers call to add the Disney Channel or drop HBO. They write letters to say that they would like to see their service add C-Span, which covers Congress, or perhaps an additional Spanish language station. They realize they have more control over the information they receive than they do with most other mass media.

Some cable operators questioned the concept of offering a basic service for a fixed rate. The basic service system provides consumers with 20 to 30 channels for one standard price. Each additional premium channels cost more. Instead, cable companies started to sell channel packages, which meant that subscribers had to choose between groups of channels and options, without the option to ungroup them.

Recent research, however, indicates that consumers are not staying in their niches. In fact, many viewers are watching the broadcast network-type programming on such services as USA Network and TNT more heavily than the narrowly focused programming designed for well-defined audiences. Convinced that consumers really do like network-type shows, TBS and other cable channels have experimented with running sitcoms in blocks when the broadcast networks are showing news or public affairs programs, so that viewers who want light-hearted entertainment can find it on cable when it isn't on the networks.

The Future of Television

We have discussed throughout this book an important trend toward demassification of mass media, and the situation could not be different with television. If network television reigned supreme from the 1950s to the 1970s as the main provider of entertainment and information to the masses, that situation started to change dramatically in the 1980s and 1990s, with the addition of cable channels, videocassettes and DVDs, video games and the Internet as providers of thousands of additional information and entertainment alternatives to an ever-growing and more demanding media audience. As we discussed in this chapter, network television went from being the main purveyor of news and entertainment—and coming close to being the only medium capable of fostering a shared political agenda for a mass audience—to being just one more voice in the wilderness competing for the attention of a shattered and fragmented audience.

This fragmentation also carried a financial price for television. As broadcast and cable networks lose revenue to other forms of media, it has become clear that television must reinvent itself in order to survive in the future. This reinvention may require altering traditional methods of programming and adapting to new technology.

Market Segmentation

Market segmentation seems to be one of the strategies devised by the networks to survive and prosper in the television business. In the 1950s, 1960s, and 1970s the networks remained unchallenged in the market, and had to design their schedules in a way that reached the widest and largest audience possible. This attitude helped to solidify the idea that the networks should always provide **general programming**—shows that included a mix of dramas, comedies, sports, news, game shows, variety, talk shows, and other genres aimed at the widest general audience possible.

With the advent of cable and satellite technologies, and the market fragmentation we have witnessed since the late 1970s, the networks began to question if their old market strategy was still sustainable. During the 1980s, the three networks resisted and delayed the necessary changes as much as they could. However, Fox emerged in the early 1990s as a viable network exactly because it dared to do things differently. Originally aiming its programming at ethnic minorities and younger audiences, it claimed niche markets obviously underserved by the traditional networks.

Despite the fact that, as Fox grew and solidified in the 1990s it also veered toward a more traditional general programming format, beefing up its sports and news divisions, for example, its initial success shook up the market, and made it clear for the three major networks that adaptation to the new reality was crucial for their survival. The market segmentation and experimentation by the networks seen in the 1990s and 2000s might be considered a direct effect of both Fox's emergence and the necessity to respond to the market segmentation introduced by cable television.

As a result, the networks have toyed with new ideas and formats since the mid-1990s. Instead of targeting the widest, broadest general audiences possible, these formats have been clearly designed to attend particular segments or niche markets, an unthinkable idea 15 years ago. Thus, CBS and ABC have alternated between trying to reach younger and older, urban and rural audiences, with short-lived programs that are dropped the next season. At least CBS's and ABC's market segmentation have paid off, as the networks enter the 2006–2007 season with high ratings. Tellingly, NBC, the network that has most resisted market segmentation and experimentation, has been struggling with dipping ratings for the past two seasons.

Television On Demand

In order to stay up-to-date with the desires of the public, broadcast networks and cable networks need to adapt the content of their programming to the match the likes and dislikes of the public and adjust program scheduling to adapt to the viewing habits of the public. With the arrival of the Internet, we have turned into an on-demand society. Many people don't wait for the morning paper or the nightly news to learn about a breaking news story, they go to the Internet to pull up the information when they demand it. Similarly, many viewers do not sit down and watch a television show during its scheduled time. Through the use of Digital Video Recorders (DVRs), viewers can record a program, shut off the set or watch a different program, and then watch the recorded program whenever they choose.

Other forms of on-demand viewing are available on the Internet. Most major networks have full episodes available for free on the networks' Web sites or for sale on Web sites such as Amazon and iTunes. While using the Internet to keep up with on-demand culture will be an important part of network television's survival, it also opens the industry up to more competition. YouTube and other video sharing Web sites offer thousands of hours of original programming that can pull viewers away from network programs that are available on television and on the Internet. To keep viewer's eyes on network television programs, networks such as CBS and BBC have partnered with YouTube to provided collaborative programming to viewers.[13]

In the future, television programs may permanently move from the television screen to the computer screen, couch potatoes may turn into mouse potatoes, and audiences may become their own programmers. But just as it has for the last 100 years, television will continue to grow and evolve with technology.

[13] Thom Patterson, "Is the Future of Television on the Web?" *CNN.com*. Retrieved from http://www.cnn.com/2008/SHOWBIZ/TV/05/01/tv.future/index.html.

Name ______________________ Course ______________ Date ________________

Chapter Questions

1. Reality television and news talk shows have become popular genres in recent years. How have they positively and negatively impacted society?

2. How is public broadcasting different from commercial broadcasting? How do these differences impact the reliability of each type of broadcasting?

3. How has the regulation of television changed over time? Should graphic images (violent, sexual, and improper content) be more federally regulated?

GREAT-GRANDAD'S LAST BATTLE

Robert Froman

When early film introduced a world of visual False Reality in the late 19th century, children, as is the case with most new technologies, were the first to adapt. Older folks, like the great-grandfather in Robert Froman's comedic memoir, struggled with reality versus the illusion created by film. That Grandad's first film contains scenes which mirror his Civil War experience only adds to the confusion, and hilarity, in the story.

One of the most memorable occasions of my youth in Caldwell, Idaho, was the Saturday afternoon I took my great-grandfather to his first picture show. I was nine or ten, and "Granddad," as he was known to his several generations of descendants, was in his 90's. The show was a "Wild Western," and that afternoon the term proved gloriously apt.

Granddad had been a more or less commissioned officer in a body of irregular Confederate troops in Missouri in 1863, and he tended to hark back to those days when he got excited or when he was in a storytelling mood. My first memory of him is of a day when my mother left me with him at his cabin while she went shopping. He was impressively gaunt, with a bony, Bible-prophet face set off by sparse, shoulder-length white hair. I was scared of him at first; but, after Mother departed, he got out his corncob pipe and sack of mullein leaf, and, as I watched, he slowly stuffed the bowl, lit it, took a couple of puffs and passed it to me. From that moment I was his to command.

"Great-Grandad's Last Battle" by Robert Froman. In *Reader's Digest*" December 1963.

On the afternoon I took him to his first movie we were a bit late. The lights had already been dimmed, and the pianist, a young music student at the local college, was running through her practice chords. I guided Granddad down the aisle to the front row. By the time he was settled, the feature was opening with Hoot Gibson riding across the range. I glanced sideways for Granddad's reaction. He was completely spellbound.

The plot thickened fast when the camera switched to a stagecoach moving toward an armed ambush. As the villains prepared to pounce, the pianist quickened the rhythm. Suddenly Granddad rose to his feet and let out a piercing rebel yell to warn the stage driver.

© sutulastock/Shutterstock.com

The pianist turned to behold, looming over her, about three feet away, a tall, tottery figure with arms flung wide, a three-day growth of white beard, and a wild look in his deep-sunk eyes. With a screech the girl leaped from bench to keyboard to piano top, and, thence to the stage, where she stumbled through the stretched sheet that served as the screen, tearing open a large flap.

Except for Granddad the audience of perhaps 40 included no one older than 12. The unscheduled hullabaloo suggested that we ought to do something, too. But what? The action on the wounded screen showed us. Riding in pursuit of the villains, Hoot jumped, fell or was shot from his horse right into the hole created by the pianist, and a friend of mine named Vergil climbed up to see what had become of him. Three or four others followed Vergil. They had just made it when the scene on the untorn part of the screen switched to a view of the villains riding off with their loot. Vergil pulled his cap pistol and let them have it.

Since the Fourth of July was approaching, nearly all of us were armed with cap pistols and well supplied with caps. Under ordinary circumstances we would never have thought of firing our guns inside the theater. But now, with the villains still on the screen, Granddad turned to his troops and called for a barrage.

The firing quickly brought the ticket-taker. In those days motion picture film was highly flammable; when the ticket taker saw the smoke from our caps, he dashed out of the theater yelling, "Fire!" When we heard his cries, we thought he was urging us on.

At last Granddad gave the command to cease firing. "No sense wasting your ammunition when they're running away!" he shouted. "Wait till they come sneaking back."

It was not long in coming. The villains reappeared on the scene, and Granddad, standing crane-tall in the front row, instructed us to prepare to fire.

At that moment half a dozen store clerks and passersby, led by the town constable, entered the theater through the back door. The piano player's hysterical description of what awaited them had inclined the members of this posse to be careful, and the darkness of the theater added to their caution. It was justified, for when the constable stuck his bead through the torn screen, he was greeted by a ragged volley of cap-pistol fire. Just then a fireman entered from the lobby, a fire-hose nozzle cradled in his arms. In the gloom he smelled smoke. "Water!" he shouted to his mates behind, and he turned the full force of it on the screen as he and his helpers dragged the hose down the left aisle. "Make way for the artillery!" Granddad yelled. "Maintain covering fire!" By this time there was not enough left of the screen to give us more than an occasional glimpse of the top of someone's head or a horse's tail. But we scarcely noticed the loss, for the drenched constable and his cohorts made fine targets. We fell back up the right aisle until we had used all our ammunition, then retreated to the street and marched off with Granddad to his cabin for a lengthy reminiscing about our battle experiences. All of us looked forward to another skirmish the next Saturday afternoon, but Granddad never again would join us at the movies. He said they were too noisy.

Name ______________________ Course ____________ Date _______________

Chapter Questions

1. In terms of his past military experience, why do you think Great-Grandad acted the way he did in the theater? How did false and immediate reality play a role in his reaction?

2. What situations have you experienced similar to Great-Granddad's? Explain.

SECTION 4 Weapons

Introduction

Since ancient times, technology has played an important role in warfare. Advancements and improvements to military weapons throughout history have forced armies to continually adopt new fighting tactics to win battles and conquer opposing armies. This is still true in the modern era where advances in drone technology, robotics, and targeting systems have led to more capable weapons that have changed the ways wars are planned and fought. This unit's readings will focus on how weapons have shaped and continue to shape society, and they show how some people look to the future with anticipation in terms of militaristic weapon advancement, while others look forward in fear and hesitation.

Ancient Warfare

Developments in warfare have, throughout history, impacted both the armies fighting with new weapons and also those fighting against them. For example, during the Iron Age, Celtic raiders were able to defeat the armies of Ancient Rome because they used stronger iron weapons, while the Roman army still used bronze. However, like many societies of ancient civilization, the Romans were able to take military aspects from their enemies to enhance their own army. By adopting the Celtic iron sword and other iron weapons, the Romans, who were much more organized than the Celts, were able to wipe them out and dominate the Mediterranean for centuries.

Impact of the Changing Ways of Warfare

Similar to Ancient civilizations, more modern societies have also been shaped by the availability and advancements in warfare and weaponry. In the eighteenth century, European nations were able to gain control of most of the world's land because of access to superior weapons. One such weapon was the machine gun, which became known as the most effective weapon of British imperial advancement. One instance of this weapon's importance occurred during the Matabele War (1893–1894), which was fought between British colonists and the Matabele tribe in modern-day Zimbabwe. In one engagement, fifty British soldiers fought off five thousand Matabele warriors with just four machine guns. Because the Matabele had never seen such a weapon, the British, who the Matabele thought were invincible because of this new weapon, were able to win the war and colonize the area, even though the British had far fewer soldiers involved.

Like the ancient Roman military, soldiers had to adapt to new weapon technology during World War I to overpower enemy troops. Trench warfare was created as a method for soldiers to hide themselves from machine gun fire. The use of trenches led to the need for a new weapon to help those in opposition of the entrenched win battles, and one such weapon was chemical gases. Ranging from disabling chemicals like tear gas to lethal agents, these were primarily used to debilitate, injure, and kill soldiers fighting in trenches. Due to the prominence and importance of chemical gases in the battles of the First World War, oftentimes it was referred to as the "chemist's war."

Warfare in Modern Society

The inventions and advancements occurring with weapons today is astonishing. Decades ago, pilotless aircraft, cyberspace warfare, and bio-hacking were only the elements of science fiction stories; however, modern militaristic technology is constantly being innovated and these ideas are being or have already been incorporated into warfare. Modern warfare technologies continue to shape the ways that wars are fought and assert how countries achieve greater power as they possess more advanced warfare technologies.

Some believe the revolutionary weapons mentioned in this unit's readings are needed to keep societies powerful; however, with any new technology, there is also a sense of hesitation and fear. Some ask whether these advancements are getting out of hand. People are left to ask if the advancement of "super weapons" helps or harms society in the long run and whether it is beneficial that advanced weapons keep societies militaristically competitive or if these weapons keep societies from finding unity.

THE RED BADGE OF COURAGE: AN EPISODE OF THE AMERICAN CIVIL WAR

Stephen Crane

Changing technology during the Civil War era rendered the weapons far more deadly and contributed to enormous casualties. During this war, many men, some as young as fifteen, went to war without battlefield experience or training with these new weapons. As a result, many soldiers were killed when their rifles were loaded incorrectly (and would not fire) either because of lack of knowledge or because, in the middle of battle, their nerves caused them to do it inaccurately. The Red Badge of Courage, *written in 1895 by Stephen Crane, is a fictional story of one soldier's first experience in battle during the Civil War. Major aspects of the story focus on the young soldier's reactions and responses to battle, the psychological impact the war has on him, and his eventual flight from the battlefield. Chapters 4 and 5 included in this text vividly describe the fear of waiting for an enemy's attack, the stress of correctly loading a musket in the middle of battle, and the dichotomy between the battlefield and the surrounding natural setting.*

This Chromolithograph by Thulstrup de Thure, which was painted in 1888, shows Civil War soldiers during the Battle of Shiloh.

Chapter 4

The brigade was halted in the fringe of a grove. The men crouched among the trees and pointed their restless guns out at the fields. They tried to look beyond the smoke.

Out of this haze they could see running men. Some shouted information and gestured as the hurried.

The men of the new regiment watched and listened eagerly, while their tongues ran on in gossip of the battle. They mouthed rumors that had flown like birds out of the unknown.

"They say Perry has been driven in with big loss."

"Yes, Carrott went t' th' hospital. He said he was sick. That smart lieutenant is commanding 'G' Company. Th' boys say they won't be under Carrott no more if they all have t' desert. They allus knew he was a—"

"Hannises' batt'ry is took."

"It ain't either. I saw Hannises' batt'ry off on th' left not more'n fifteen minutes ago."

"Well—"

"Th' general, he ses he is goin' t' take th' hull command of th' 304th when we go inteh action, an' then he ses we'll do sech fightin' as never another one reg'ment done."

"They say we're catchin' it over on th' left. They say th' enemy driv' our line inteh a devil of a swamp an' took Hannises' batt'ry."

"No sech thing. Hannises' batt'ry was 'long here 'bout a minute ago."

"That young Hasbrouck, he makes a good off'cer. He ain't afraid 'a nothin'."

"I met one of th' 148th Maine boys an' he ses his brigade fit th' hull rebel army fer four hours over on th' turnpike road an' killed about five thousand of 'em. He ses one more sech fight as that an' th' war 'll be over."

"Bill wasn't scared either. No, sir! It wasn't that. Bill ain't a-gittin' scared easy. He was jest mad, that's what he was. When that feller trod on his hand, he up an' sed that he was willin' t' give his hand t' his country, but he be dumbed if he was goin' t' have every dumb bushwhacker in th' kentry walkin' 'round on it. So he went t' th' hospital disregardless of th' fight. Three fingers was crunched. Th' dern doctor wanted t' amputate 'm, an' Bill, he raised a heluva row, I hear. He's a funny feller."

The din in front swelled to a tremendous chorus. The youth and his fellows were frozen to silence. They could see a flag that tossed in the smoke angrily. Near it were the blurred and agitated forms of troops. There came a turbulent stream of men across the fields. A battery changing position at a frantic gallop scattered the stragglers right and left.

A shell screaming like a storm banshee went over the huddled heads of the reserves. It landed in the grove, and exploding redly flung the brown earth. There was a little shower of pine needles.

Bullets began to whistle among the branches and nip at the trees. Twigs and leaves came sailing down. It was as if a thousand axes, wee and invisible, were being wielded. Many of the men were constantly dodging and ducking their heads.

The lieutenant of the youth's company was shot in the hand. He began to swear so wondrously that a nervous laugh went along the regimental line. The officer's profanity sounded conventional. It relieved the tightened senses of the new men. It was as if he had hit his fingers with a tack hammer at home.

He held the wounded member carefully away from his side so that the blood would not drip upon his trousers.

The captain of the company, tucking his sword under his arm, produced a handkerchief and began to bind with it the lieutenant's wound. And they disputed as to how the binding should be done.

The battle flag in the distance jerked about madly. It seemed to be struggling to free itself from an agony. The billowing smoke was filled with horizontal flashes.

Men rushing swiftly emerged from it. They grew in numbers until it was seen that the whole command was fleeing. The flag suddenly sank down as if dying. Its motion as it fell was a gesture of despair.

Wild yells came from behind the walls of smoke. A sketch in gray and red dissolved into a moblike body of men who galloped like wild horses. The veteran regiments on the right and left of the 304th immediately began to jeer. With the passionate song of the bullets and the banshee shrieks of shells were mingled loud catcalls and bits of facetious advice concerning places of safety.

But the new regiment was breathless with horror. "Gawd! Saunders's got crushed!" whispered the man at the youth's elbow. They shrank back and crouched as if compelled to await a flood.

The youth shot a swift glance along the blue ranks of the regiment. The profiles were motionless, carven; and afterward he remembered that the color sergeant was standing with his legs apart, as if he expected to be pushed to the ground.

The following throng went whirling around the flank. Here and there were officers carried along on the stream like exasperated chips. They were striking about them with their swords and with their left fists, punching every head they could reach. They cursed like highwaymen.

A mounted officer displayed the furious anger of a spoiled child. He raged with his head, his arms, and his legs.

Another, the commander of the brigade, was galloping about bawling. His hat was gone and his clothes were awry. He resembled a man who has come from bed to go to a fire. The hoofs of his horse often threatened the heads of the running men, but they scampered with singular fortune. In this rush they were apparently all deaf and blind. They heeded not the largest and longest of the oaths that were thrown at them from all directions.

Frequently over this tumult could be heard the grim jokes of the critical veterans; but the retreating men apparently were not even conscious of the presence of an audience.

The battle reflection that shone for an instant in the faces on the mad current made the youth feel that forceful hands from heaven would not have been able to have held him in place if he could have got intelligent control of his legs.

There was an appalling imprint upon these faces. The struggle in the smoke had pictured an exaggeration of itself on the bleached cheeks and in the eyes wild with one desire.

The sight of this stampede exerted a floodlike force that seemed able to drag sticks and stones and men from the ground. They of the reserves had to hold on. They grew pale and firm, and red and quaking.

The youth achieved one little thought in the midst of this chaos. The composite monster which had caused the other troops to flee had not then appeared. He resolved to get a view of it, and then, he thought he might very likely run better than the best of them.

Chapter 5

This picture shows Union soldiers in trenches before the Battle of Petersburg, Virginia on June 9, 1864.

There were moments of waiting. The youth thought of the village street at home before the arrival of the circus parade on a day in the spring. He remembered how he had stood, a small, thrillful boy, prepared to follow the dingy lady upon the white horse, or the band in its faded chariot. He saw the yellow road, the lines of expectant people, and the sober houses. He particularly remembered an old fellow who used to sit upon a cracker box in front of the store and feign to despise such exhibitions. A thousand details of color and form surged in his mind. The old fellow upon the cracker box appeared in middle prominence.

Some one cried, "Here they come!"

There was rustling and muttering among the men. They displayed a feverish desire to have every possible cartridge ready to their hands. The boxes were pulled around into various positions, and adjusted with great care. It was as if seven hundred new bonnets were being tried on.

The tall soldier, having prepared his rifle, produced a red handkerchief of some kind. He was engaged in knotting it about his throat with exquisite attention to its position, when the cry was repeated up and down the line in a muffled roar of sound.

"Here they come! Here they come!" Gun locks clicked.

Across the smoke-infested fields came a brown swarm of running men who were giving shrill yells. They came on, stooping and swinging their rifles at all angles. A flag, tilted forward, sped near the front.

As he caught sight of them the youth was momentarily startled by a thought that perhaps his gun was not loaded. He stood trying to rally his faltering intellect so that he might recollect the moment when he had loaded, but he could not.

A hatless general pulled his dripping horse to a stand near the colonel of the 304th. He shook his fist in the other's face. "You've got to hold 'em back!" he shouted, savagely; "you've got to hold 'em back!"

In his agitation the colonel began to stammer. "A-all r-right, General, all right, by Gawd! We-we 'll do our—we-we 'll d-d-do-do our best, General." The general made a passionate gesture and galloped away. The colonel, perchance to relieve his feelings, began to scold like a wet parrot. The youth, turning swiftly to make sure that the rear was unmolested, saw the commander regarding his men in a highly resentful manner, as if he regretted above everything his association with them.

The man at the youth's elbow was mumbling, as if to himself: "Oh, we 're in for it now! oh, we 're in for it now!"

The captain of the company had been pacing excitedly to and fro in the rear. He coaxed in schoolmistress fashion, as to a congregation of boys with primers. His talk was an endless repetition. "Reserve your fire, boys—don't shoot till I tell you—save your fire—wait till they get close up—don't be damned fools—"

Perspiration streamed down the youth's face, which was soiled like that of a weeping urchin. He frequently, with a nervous movement, wiped his eyes with his coat sleeve. His mouth was still a little ways open.

He got the one glance at the foe-swarming field in front of him, and instantly ceased to debate the question of his piece being loaded.

Before he was ready to begin—before he had announced to himself that he was about to fight—he threw the obedient well-balanced rifle into position and fired a first wild shot. Directly he was working at his weapon like an automatic affair.

He suddenly lost concern for himself, and forgot to look at a menacing fate. He became not a man but a member. He felt that something of which he was a part—a regiment, an army, a cause, or a country—was in crisis. He was welded into a common personality which was dominated by a single desire. For some moments he could not flee no more than a little finger can commit a revolution from a hand.

If he had thought the regiment was about to be annihilated perhaps he could have amputated himself from it. But its noise gave him assurance. The regiment was like a firework that, once ignited, proceeds superior to circumstances until its blazing vitality fades. It wheezed and banged with a mighty power. He pictured the ground before it as strewn with the discomfited.

There was a consciousness always of the presence of his comrades about him. He felt the subtle battle brotherhood more potent even than the cause for which they were fighting. It was a mysterious fraternity born of the smoke and danger of death.

He was at a task. He was like a carpenter who has made many boxes, making still another box, only there was furious haste in his movements. He, in his thoughts, was careering off in other places, even as the carpenter who as he works whistles and thinks of his friend or his enemy, his home or a saloon. And these jolted dreams were never perfect to him afterward, but remained a mass of blurred shapes.

Presently he began to feel the effects of the war atmosphere—a blistering sweat, a sensation that his eyeballs were about to crack like hot stones. A burning roar filled his ears.

Following this came a red rage. He developed the acute exasperation of a pestered animal, a well-meaning cow worried by dogs. He had a mad feeling against his rifle, which could only be used against one life at a time. He wished to rush forward and strangle with his fingers. He craved a power that would enable him to make a world-sweeping gesture and brush all back. His impotency appeared to him, and made his rage into that of a driven beast.

Buried in the smoke of many rifles his anger was directed not so much against the men whom he knew were rushing toward him as against the swirling battle phantoms which were choking him, stuffing their smoke robes down his parched throat. He fought frantically for respite for his senses, for air, as a babe being smothered attacks the deadly blankets.

There was a blare of heated rage mingled with a certain expression of intentness on all faces. Many of the men were making low-toned noises with their mouths, and these subdued cheers, snarls, imprecations, prayers, made a wild, barbaric song that went as an undercurrent of sound, strange and chantlike with the resounding chords of the war march. The man at the youth's elbow was babbling. In it there was something soft and tender like the monologue of a babe. The tall soldier was swearing in a loud voice. From his lips came a black procession of curious oaths. Of a sudden another broke out in a querulous way like a man who has mislaid his hat. "Well, why don't they support us? Why don't they send supports? Do they think—"

The youth in his battle sleep heard this as one who dozes hears.

There was a singular absence of heroic poses. The men bending and surging in their haste and rage were in every impossible attitude. The steel ramrods clanked and clanged with incessant din as the men pounded them furiously into the hot rifle barrels. The flaps of the cartridge boxes were all unfastened, and bobbed idiotically with each movement. The rifles, once loaded, were jerked to the shoulder and fired without apparent aim into the smoke or at one of the blurred and shifting forms which upon the field before the regiment had been growing larger and larger like puppets under a magician's hand.

The officers, at their intervals, rearward, neglected to stand in picturesque attitudes. They were bobbing to and fro roaring directions and encouragements. The dimensions of their howls were extraordinary. They expended their lungs with prodigal wills. And often they nearly stood upon their heads in their anxiety to observe the enemy on the other side of the tumbling smoke.

The lieutenant of the youth's company had encountered a soldier who had fled screaming at the first volley of his comrades. Behind the lines these two were acting a little isolated scene. The man was blubbering and staring with sheeplike eyes at the lieutenant, who had seized him by the collar and was pommeling him. He drove him back into the ranks with many blows. The soldier went mechanically, dully, with his animal-like eyes upon the officer. Perhaps there was to him a divinity expressed in the voice of the other—stern, hard, with no reflection of fear in it. He tried to reload his gun, but his shaking hands prevented. The lieutenant was obliged to assist him.

The men dropped here and there like bundles. The captain of the youth's company had been killed in an early part of the action. His body lay stretched out in the position of a tired man resting, but upon his face there was an astonished and sorrowful look, as if he thought some friend had done him an ill turn. The babbling man was grazed by a shot that made the blood stream widely down his face. He clapped both hand to his head. "Oh!" he said, and ran. Another grunted suddenly as if he had been struck by a club in the stomach. He sat down and gazed ruefully. In his eyes there was mute, indefinite reproach. Farther up the line a man, standing behind a tree, had had his knee joint splintered by a ball. Immediately he had dropped his rifle and gripped the tree with both arms. And there he remained, clinging desperately and crying for assistance that he might withdraw his hold upon the tree.

At last an exultant yell went along the quivering line. The firing dwindled from an uproar to a last vindictive popping. As the smoke slowly eddied away, the youth saw that the charge had been repulsed. The enemy were scattered into reluctant groups. He saw a man climb to the top of the fence, straddle the rail, and fire a parting shot. The waves had receded, leaving bits of dark "debris" upon the ground.

Some in the regiment began to whoop frenziedly. Many were silent.

Apparently they were trying to contemplate themselves.

After the fever had left his veins, the youth thought that at last he was going to suffocate. He became aware of the foul atmosphere in which he had been struggling. He was grimy and dripping like a laborer in a foundry. He grasped his canteen and took a long swallow of the warmed water.

A sentence with variations went up and down the line. "Well, we 've helt 'em back. We 've helt 'em back; derned if we haven't." The men said it blissfully, leering at each other with dirty smiles.

The youth turned to look behind him and off to the right and off to the left. He experienced the joy of a man who at last finds leisure in which to look about him.

Under foot there were a few ghastly forms motionless. They lay twisted in fantastic contortions. Arms were bent and heads were turned in incredible ways. It seemed that the dead men must have fallen from some great height to get into such positions. They looked to be dumped out upon the ground from the sky.

From a position in the rear of the grove a battery was throwing shells over it. The flash of the guns startled the youth at first. He thought they were aimed directly at him. Through the trees he watched the black figures of the gunners as they worked swiftly and intently. Their labor seemed a complicated thing. He wondered how they could remember its formula in the midst of confusion.

The guns squatted in a row like savage chiefs. They argued with abrupt violence. It was a grim pow-wow. Their busy servants ran hither and thither.

A small procession of wounded men were going drearily toward the rear. It was a flow of blood from the torn body of the brigade.

To the right and to the left were the dark lines of other troops. Far in front he thought he could see lighter masses protruding in points from the forest. They were suggestive of unnumbered thousands.

Once he saw a tiny battery go dashing along the line of the horizon. The tiny riders were beating the tiny horses.

From a sloping hill came the sound of cheerings and clashes. Smoke welled slowly through the leaves.

Batteries were speaking with thunderous oratorical effort. Here and there were flags, the red in the stripes dominating. They splashed bits of warm color upon the dark lines of troops.

The youth felt the old thrill at the sight of the emblems. They were like beautiful birds strangely undaunted in a storm.

As he listened to the din from the hillside, to a deep pulsating thunder that came from afar to the left, and to the lesser clamors which came from many directions, it occurred to him that they were fighting, too, over there, and over there, and over there. Heretofore he had supposed that all the battle was directly under his nose.

As he gazed around him the youth felt a flash of astonishment at the blue, pure sky and the sun gleamings on the trees and fields. It was surprising that Nature had gone tranquilly on with her golden process in the midst of so much devilment.

Name ______________________ Course ____________ Date ______________

Chapter Questions

1. This reading discusses a battle during the Civil War. One soldier is described as "working at his weapon like an automatic affair. He suddenly lost concern for himself and, forgot to look at a menacing fate; he became not a man but a member." Discuss this quote. How does Crane describe the soldier as he puts together his weapon?

2. Throughout his novel, Crane discusses nature as opposed to warfare and technology. Discuss the quote, "It was surprising that Nature had gone tranquilly on with her golden process in the midst of so much devilment." How does Crane characterize war and weapons through statements like this one? How does he characterize nature?

OVERVIEW OF WORLD WAR I AND WORLD WAR II WEAPONS

From Historical Context and Literature

Hubert P. van Tuyll

Everett Historical/Shutterstock.com

This picture was taken during the last month of World War I, and it shows French tanks passing through Rampont, France. The armored vehicles were introduced to break the deadlock of the trench warfare of the Western Front in the final stages of war.

Introduction (World War I)

The end of the nineteenth century was a period of tremendous change all over the world. The "Second Industrial Revolution" brought even more rapid and far-reaching innovations than had the first phase of industrialization, including the perfection of the steelmaking process, the harnessing of electricity, and the invention of the automobile. Cities expanded and economies surged, bringing more and more wealth into the middle class but also steadily raising living standards for the lower classes as well. This was the Gilded Age, the Belle Époque, an era of wealth, progress, and self-satisfied optimism. But this optimism was tinged with anxiety. New scientific theories and discoveries, like Albert Einstein's theory of relativity and Sigmund Freud's discovery of the unconscious, cast doubt on the very foundations of Enlightenment beliefs about humankind and progress. Women, the working class, colonized peoples in Asia, Africa, and the Caribbean, and other social groups traditionally excluded from power organized into mass movements and demanded the extension of civil rights. Politics, once the dignified preserve of the elite, became increasingly dominated by the raucous politics of the masses. And nationalism and racism became more overt and dominant in public life. Internationally, the competition for markets, resources and prestige grew ever more fierce, as the Western powers laid imperialistic claim to the last portions of the planet. Newcomer states to the international scene, like Germany, Italy, Japan, and the United States, were particularly keen to stake their claims to Great Power status. Tensions rose and skirmishes increased, until finally, in 1914, the general European war that so many had both feared and hoped for finally broke out. The Great War, it was widely believed by all kinds of social groups, would change everything. And it did, though not entirely in the ways they had expected. Vast and important change came from the war, much of it even positive, but it came at a terrible price. And in the end its legacy was a society even more anxious, doubtful, and disillusioned.

Weapons of Mass Destruction

The war began very much in the spirit of the nineteenth century wars, with rapid frontal attacks aimed at capturing and holding territory. That tactic, and the belief in a quick victory shared by all sides, was soon confounded by the new reality of industrial warfare. The scientific and mechanical advancements of the Industrial Revolution had led to the development of a whole new generation of frightfully lethal weapons. Machine guns, barbed wire, tanks, flamethrowers, chemical weapons, airplanes, and submarines were all either first introduced or first used extensively in the First World War. Artillery had been used for hundreds of years, of course, but by 1914 the size and accuracy of the guns and the explosive power of the shells had increased tremendously and continued to increase throughout the war. The Germans deployed a number of enormous 419-mm guns, nicknamed "Big Berthas," which could fire high-explosive shells weighing nearly 2,000 pounds for distances of 75 miles (Keegan, 1999, p. 406). The weapon that most terrified the ordinary soldier, however, was poison gas. Tear gas, the first to be used at the very beginning of the war, had little practical effect. But the Germans, followed by the French, British and, eventually, Americans, quickly devised more effective and lethal types of chemical weapons. Chlorine and phosgene gases made the lungs fill with fluid, causing the soldier to literally drown on dry land. Mustard gas caused blindness, painful skin blisters, bleeding, and scorched windpipes so that breathing was painful and difficult. Mustard gas didn't kill as efficiently as chlorine and phosgene, but its effects were arguably more frightening. Its victims could take weeks to die, in utter agony. And a great many of those subject to mustard gas attacks survived to live out the rest of their lives blind or unable to speak above a hoarse whisper. The effect of all these new weapons on the soldiers fighting the war was confusion, demoralization, and paralysis. In the Battle of the Marne, which stopped the German advance toward Paris in September 1914, more than two million men were thrown into the fight on all sides, and more than 500,000 were killed or wounded. Other battles were even more gruesome. At the Battle of the Somme in July 1916, British forces pounded the German trenches for an entire week with heavy artillery, firing 1,627,824 shells along a 10-mile front. When the barrage ceased, 100,000 British soldiers started forward in a traditional frontal charge toward the German lines. The surviving German soldiers literally mowed down the advancing ranks of British "Tommies" with machine-gun fire. In the first hour of the battle, the British lost 30,000 men—killed, wounded, or captured. At the end of the day, 19,240 British were dead, and more than 38,000 wounded or captured. The battle dragged on for another four and a half months. At the end of the engagement, only a few miles had been gained by the Allies. The cost was over 619,000 Allied casualties (killed, wounded, and captured) and a similar number for the Germans (Neiberg, 2005, pp. 192–202).

This picture shows a British World War I machine gun crew in a front line trench.

Trench Warfare

The advancement of weaponry had outpaced military tactics. Having no practical experience defending against the new weapons, the commanders simply ordered the troops to dig in to the ground. An elaborate network of fortified trenches, some with subterranean rooms, was erected by both sides on the Western front, stretching from the English Channel to the French border with Switzerland. A similar network was constructed on the Italian and Austro-Hungarian lines high in the Alps. From these positions, soldiers took turns periodically leaping out of their trenches and rushing across the few hundred yards of no man's land to attempt to take the other side's trenches, generally with similar results as at the Somme. Although occasionally some ground was won and new trenches built by one side or another, for the most part the war from early 1915 to early 1917 was a stalemate, with neither side gaining much of a strategic advantage. For the commanders, this hybrid of the nineteenth-century tactics and twentieth-century industrial weapons was intensely frustrating and lacking in chivalry. During the gloomiest period of the war, in 1916, Lord Herbert Kitchener, commander of all British forces, threw his hands up in despair, exclaiming, "I don't know what is to be done. This isn't war." The sentiment was echoed by a rank-and-file British soldier, who wrote back home to India, "This is not war. It is the ending of the world" (Hochschild, 2011, pp. 142, xiv). Life for the ordinary soldiers was a routine of profound boredom and discomfort, punctuated by periods of sheer terror during artillery barrages and infantry assaults. Conditions in the trenches, where many soldiers spent months and even years without a break, were cramped and uncomfortable, hot and stifling in the summer, cold and damp in the winter, and filled with knee-deep mud when it rained. And always there were rats—eating supplies, feasting on the corpses of comrades, carrying fleas and disease. The patriotic enthusiasm that had characterized so many young men's volunteering for the war in 1914 had quickly given way to a sullen hatred of the generals who seemed to have no idea what to do and a desperate determination to survive by doing as little as they could get away with. By 1917, desertions in the tens of thousands were affecting every army. Some units mutinied en masse, refusing orders to advance, sometimes even shooting their officers.

The Home Fronts

But the First World War was a different sort of war experience not only for the soldiers. The unprecedented requirements in weapons, vehicles, uniforms, bandages, and provisions for the enormous numbers of men in the field meant that much of the economy had to be shifted from civilian to war production. In every belligerent country, government oversight bodies were established to coordinate the production of war material, and they often intervened directly in the civilian economy, setting production quotas, prices, and wages and criminalizing strikes. The severe labor shortages caused by the massive conscription of men into the military was dealt with by the recruitment, and sometimes conscription, of women and, in some of the harder-hit countries like Germany, increasingly young boys to take the men's places in the factories. The government boards in most countries also instituted rationing of basic foodstuffs like sugar, potatoes, and meat to prevent panic buying and to ensure a steady supply for the troops. This was a war; all sides were told, not simply of one government against another, but rather of entire societies against one another. The idea of the home front, popularized for the first time during the First World War, conveyed the idea that everyone was part of the war effort, and that the front was just as much in the factories and homes at the rear as in the trenches.

Introduction (World War I)

This picture shows the wrecked framework of Museum of Science and Industry in Nagasaki, Japan after the atomic bomb was dropped on the city on August 6, 1945. World War II.

World War I left massive instability in its wake. The Versailles treaty could not cope with an environment that included the disintegration of two empires, the collapse of four monarchies, and financial and human losses that left the major European imperial states crippled. The wave of radicalism at war's end had seemingly been contained, but it had left Russia a communist state and therefore isolated from other countries. A series of economic crises growing out of the war culminated in the Great Depression. Poverty and instability left the people vulnerable to the appeals of radical demagoguery, especially in Italy and Germany, two countries embittered by Versailles—albeit for very different reasons. Eventually, radical parties determined to upset the international status quo seized power in both countries, helped by the unpopularity and weakness of their governments. In Japan also there existed militant factions dissatisfied with the world order, and these would push their country to obtain an empire of its own. Ties between these three countries began to strengthen in 1935, and eventually they formed the losing alliance in World War II: the Axis. The war they unleashed would be the bloodiest in history, costing between 55 and 75 million lives—most of whom were civilians. The greatest loss of life took place in Eastern Europe, but millions also died in China. Ruthless warfare, aerial bombardment, and scorched-earth policies added incalculable physical destruction to the war's immediate results.

Atom Bomb

Very few sane people today debate the atrocity of the Holocaust. The same is not true of the atomic bombing of Japan, which remains controversial. The Pacific war became extraordinarily bitter, with genuine hatred on both sides, and Americans dreaded the thought of invading Japan, especially after the spectacular bloodbaths at Okinawa, Iwo Jima, Tarawa, and Peleliu. The United States hoped to subdue Japan with bombing, and this was implemented in 1945 with massive conventional raids. Japan's cities were far more vulnerable to bombing than Germany's because most of the buildings were made of wood. The conventional bombing of Tokyo alone cost Japan over 100,000 lives. But Japan did not surrender. Whether it was about to surrender remains a controversy in diplomatic history. However, the issue became moot on August 6, 1945, when a single atomic bomb was dropped on Hiroshima. The explosion obliterated more than half of the city, damaged structures for five square miles, and killed 80,000 people; half again that number died in the following year from the after effects. A second bomb, intended to convince the Japanese that the United States had a limitless supply of them, was dropped on Nagasaki three days later. The human toll was about half that of Hiroshima, but it was enough. The two bombs and the Soviet declaration of war convinced the emperor that resistance was futile, and he ordered the August 15 surrender.

The Minds behind the Bomb

The atom bomb was developed in secret in the United States, though many of the chief scientists and innovators were Europeans, some of whom knew about scientific advances under way in Germany. Two physicists, Leo Szilard (1898–1964), a Jewish Hungarian, and Enrico Fermi (1901–1954), an Italian, created a plan to produce a nuclear reactor. They brought their work to the United States government and told them of its potential application to weaponry. The scientists urged Albert Einstein to use his fame to help their cause. Einstein wrote a letter to President Roosevelt explaining that Germany was on the brink of harnessing this technology, and it was imperative that another nation invest in the project. The Manhattan district of the U.S. Army Corps of Engineers organized laboratories, engineering, and construction plants. The Manhattan Project was top secret, although it involved hundreds of scientists, civilians, and military in production centers at Oak Ridge, Tennessee, and Hanford, Washington. It was directed by army brigadier general Leslie R. Groves (1896–1970), who also supervised the building of the Pentagon. Scientists from Great Britain and Canada also assisted, making it a joint Allied endeavor. When J. Robert Oppenheimer (1904–1967) joined the project, he suggested a laboratory be located in the remote New Mexican mountains at Los Alamos. When sufficient investigations were complete, they set out for the even remoter desert of Alamogordo to test the bomb in a secret operation known as Trinity. The bomb erupted in a bright flash of light, heat, and a wave of wind. A visually striking mushroom-shaped cloud and a fireball rose above the earth some 40,000 feet. Secretary of War Henry Lewis Stimson (1867–1950) knew that using the bomb would not only devastate the city it hit but also shock the entire world. Its destructive power would give the United States new leverage in global affairs. Many Manhattan Project scientists, including Szilard, appealed to President Harry S. Truman (1884–1972) not to use the bomb. The massive destruction of lives, they argued, was immoral. Truman, however, ordered the use of the bomb; he would later argue that its use had actually saved lives by eliminating the need for an invasion of the Japanese mainland, although this claim remains controversial.

Name ______________________ Course ____________ Date _______________

Chapter Questions

1. Numerous new and advanced weapons and strategies were used in WWI and WWII, including machine guns, tanks, flame throwers, chemical weapons, and trench warfare. Research the history of one of these weapons and discuss the findings with the class.

2. Controversy remains, even decades later, about the dropping of the atomic bombs by Allied Forces on Hiroshima and Nagasaki. Research both sides of this controversial topic and discuss these viewpoints with your classmates.

DULCE ET DECORUM EST

Wilfred Owen

World War I saw either the introduction or first extensive use of technologies that made killing more brutal and more efficient. Any honor, glory and heroism associated with previous wars disappeared in a hail of machine gun bullets, staggering explosions, or a fog of poison gas hurled by an enemy soldiers rarely saw face to face. British soldier-poet Wilfred Owen composed "Dulce et Decorum Est" to illustrate the horrors of poisoned gas, employed by all sides despite a pre-war agreement banning its use. In the years after the war "Dulce et Decorum Est" was recognized as one of the great poems of the era. Owen never knew this as he was killed in action November 4, 1918, one week before an Armistice ended the fighting. The Latin phrase that ends the poem is from the Roman poet Horace and is translated, "It is sweet and fitting to die for one's country."

Bent double, like old beggars under sacks,
Knock-kneed, coughing like hags, we cursed through sludge,
Till on the haunting flares we turned our backs,
And towards our distant rest began to trudge.
Men marched asleep. Many had lost their boots,
But limped on, blood-shod. All went lame; all blind;
Drunk with fatigue; deaf even to the hoots
Of gas-shells dropping softly behind.

Dulce et Decorum Est, Wilfred Owen, published in *Poems* in 1921 by Viking Press.

Gas! GAS! Quick, boys!—An ecstasy of fumbling
Fitting the clumsy helmets just in time,
But someone still was yelling out and stumbling
And flound'ring like a man in fire or lime.—
Dim through the misty panes and thick green light,
As under a green sea, I saw him drowning.

In all my dreams before my helpless sight,
He plunges at me, guttering, choking, drowning.

If in some smothering dreams, you too could pace
Behind the wagon that we flung him in,
And watch the white eyes writhing in his face,
His hanging face, like a devil's sick of sin;
If you could hear, at every jolt, the blood
Come gargling from the froth-corrupted lungs,
Obscene as cancer, bitter as the cud
Of vile, incurable sores on innocent tongues,—
My friend, you would not tell with such high zest
To children ardent for some desperate glory,
The old Lie: *Dulce et decorum est*
Pro patria mori.

Name ______________________ Course ______________ Date ________________

Chapter Questions

1. Though there is a death by gas in the poem what evidence is there that gas was used in World War I more as a weapon of intimidation than destruction?

2. Imagine that you were a soldier in the midst of a gas attack. Describe your emotions. What actions might you take?

THE UNBLINKING STARE: THE DRONE WAR IN PAKISTAN

Steve Coll

Steve Coll, a Pulitzer Prize winning journalist and dean of the Graduate School of Journalism at Columbia University, challenges the American government's assertion that armed drones represent a safer, more precise, and ultimately humanitarian air combat capability. Citing evidence from the air war over Pakistan, 2008–2013, Coll argues that drones are not nearly as exact as advocates claim, and that their prolonged use in sensitive regions may not serve the best interests of the United States. Paying particular attention to noncombatant deaths and the attitudes of populations affected by the air war, he contends that the costs associated with drone operations could dwarf any military benefits that accrue during these secretive, undeclared wars on civilian populations. In what comes across as a scathing indictment of official and unofficial government policy, Coll adds to the steady din of criticism that surrounds the application of a weapons system that promises to change warfare in ways comparable to the Macedonian phalanx, the English longbow and Chinese gunpowder. Drones provide an ability to find, fix and ultimately liquidate targets—human and otherwise—from great distances and in complete safety, and it is this last element that, ironically, renders them dangerous from moral and political standpoints. Governments that might hesitate to commit human resources to risky ventures now have fewer reasons to exercise restraint, and war, which was once a last resort, now becomes a more viable and attractive option.

This picture shows the Predator drone taking off from Balad Air Base Iraq in 2008.

At the Pearl Continental Hotel, in Peshawar, a concrete tower enveloped by flowering gardens, the management has adopted security precautions that have become common in Pakistan's upscale hospitality industry: razor wire, vehicle barricades, and police crouching in bunkers, and fingering machine guns. In June, on a hot weekday morning, Noor Behram arrived at the gate carrying a white plastic shopping bag full of photographs. He had a four-inch black beard and wore a blue shalwarkameez and a flat Chitrali hat. He met me in the lobby. We sat down, and Behram spilled his photos onto a table. Some of the prints were curled and faded. For the past seven years, he said, he has driven around North Waziristan on a small red Honda motorcycle, visiting the sites of American drone missile strikes as soon after an attack as possible.

Behram is a journalist from North Waziristan, in northwestern Pakistan, and also works as a private investigator. He has been documenting the drone attacks for the Foundation for Fundamental Rights, a Pakistani nonprofit that is seeking redress for civilian casualties. In the beginning, he said, he had no training and only a cheap camera. I picked up a photo that showed Behram outdoors, in a mountainous area, holding up a shredded piece of women's underwear. He said it was taken during his first investigation, in June, 2007, after an aerial attack on a training camp. American and Pakistani newspapers reported at the time that drone missiles had killed Al Qaeda-linked militants. There were women nearby as well. Although he was unable to photograph the victims' bodies, he said, "I found charred, torn women's clothing—that was the evidence."

Since then, he went on, he has photographed about a hundred other sites in North Waziristan, creating a partial record of the dead, the wounded, and their detritus. Many of the faces before us were young. Behram said he learned from conversations with editors and other journalists that if a drone missile killed an innocent adult male civilian, such as a vegetable vender or a fruit seller, the victim's long hair and beard would be enough to stereotype him as a militant. So he decided to focus on children.

Many of the prints had dates scrawled on the back. I looked at one from September 10, 2010. It showed a bandaged boy weeping; he appeared to be about seven years old. There was a photo of a girl with a badly broken arm, and one of another boy, also in tears, apparently sitting in a hospital. A print from August 23, 2010, showed a dead boy of perhaps ten, the son of an Afghan refugee named Bismillah Khan, who lived near a compound associated with the Taliban fighting group known as the Haqqani network. The boy's skull had been bashed in.

This picture shows the MQ-1 Predator drone being rolled into position.

Armed drones are slow-moving pilotless aircraft equipped with cameras, listening devices, and air-to-ground missiles. They can hover over their targets for hours, transmitting video feed of the scene below, and then strike suddenly. They can be flown by remote control from great distances. The models used by the C.I.A.—the Predator and the Reaper—look like giant robotic flying bugs, with unusual flaps and pole-like protrusions. Pilots steer them and fire missiles while sitting before video monitors; during a C.I.A. mission over Pakistan, a pilot might be at a base in Jalalabad, Afghanistan, or as far away as Nevada.

Last year, in a speech at the National Defense University, President Obama acknowledged that American drones had killed civilians. He called these incidents "heartbreaking tragedies," which would haunt him and those in his chain of command for "as long as we live." But he went on to defend drones as the most discriminating aerial bombers available in modern warfare—preferable to piloted aircraft or cruise missiles. Jets and missiles cannot linger to identify and avoid noncombatants before striking, and, the President said, they are likely to cause "more civilian casualties and more local outrage."

The President's commitment to what his Administration calls "surgical strikes" against terrorists and guerrillas has come to define his approach to war and counterterrorism. The decision to enter into a conflict with the Islamic State in Iraq and al-Sham, or ISIS, in which a precision air war, including the use of drones, figures heavily, is the latest, and perhaps the riskiest, manifestation of a growing reliance on targeted air strikes to manage terrorist threats. The strategy against ISIS is derived from the President's experience commanding the C.I.A.'s drone war in Pakistan, and from similar but less active drone campaigns in Yemen and Somalia.

The conflict with ISIS once again pits the world's most technologically advanced military against a stateless guerrilla group. In such a contest, civilian casualties are not only a moral issue; they constitute a front in a social-media contest over justice and credibility. This summer, when the Administration opened its air assault in northern Iraq, ISIS media specialists tweeted photos of children who reportedly had been killed and wounded earlier by American drones in Yemen.

Obama's advocacy of drones has widespread support in Washington's foreign-policy and defense establishments. As Secretary of State, Hillary Clinton wholeheartedly backed the drone campaigns in Pakistan and Yemen. Republican hawks like John McCain and Lindsey Graham, who otherwise criticize the President as effete and indecisive, are also enthusiastic. But do drones actually represent a humanitarian advance in air combat? Or do they create a false impression of exactitude? And do they really serve the best interests of the United States?

Pakistan has absorbed more drone strikes—some four hundred—than any other country, and has been a test bed for the Administration's hypotheses about the future of American airpower. Between mid-2008 and mid-2013, C.I.A.-operated drones waged what amounted to an undeclared, remotely controlled air war over North and South Waziristan, a sparse borderland populated almost entirely by ethnic Pashtuns. As the campaign evolved, it developed a dual purpose: to weaken Al Qaeda, and to suppress Taliban fighters who sought to cross into Afghanistan to attack American troops after Obama ordered a "surge" of forces there, in December, 2009. (Drone strikes continue in Pakistan; seventeen have been reported so far this year.) The drone war in Pakistan took place during an increasingly toxic, mutually resentful period in the long, unhappy chronicle of relations between the United States and Pakistan. To many Pakistanis, including Army officers and intelligence officials in the Inter-Services Intelligence Directorate, or I.S.I., drone strikes have symbolized American arrogance. Within the C.I.A. and the White House, a belief took hold that Pakistani generals and intelligence chiefs were unreliable partners in the fight against Al Qaeda and the

Taliban. Administration officials concluded that since Pakistan wouldn't help adequately to protect U.S. soldiers and American cities, they would send drones to do the job.

President Obama and C.I.A. officials characterize the drone campaign as a major success, because it significantly reduced the ability of "core Al Qaeda"—the organization founded by Osama bin Laden—to carry out terrorism on American soil and against American and British aviation. Moreover, Obama's advisers argue, the drone war achieved this while inflicting few civilian casualties. And, they say, the machines helped the United States avoid conventional bombing or ground raids in Pakistan, which would have put American troops at risk and created even greater chaos and anger in an unstable country that possesses more than a hundred nuclear weapons.

In a 2012 report that was based on nine months of data analysis and field interviews, a team of law students from New York University and Stanford concluded that the dominant narrative in the U.S. about the use of drones in Pakistan—"a surgically precise and effective tool that makes the United States safer by enabling 'targeted killing' of terrorists, with minimal downsides or collateral impacts"—is false. The researchers found that C.I.A.-operated drones were nowhere near as discriminating toward noncombatants as the agency's leaders have claimed. Various estimates have put the civilian death toll in the hundreds. An analysis of media reports by the New America Foundation concluded that drones probably killed some two hundred and fifty to three hundred civilians in the decade leading up to 2014. Researchers working under Chris Woods at the London-based Bureau of Investigative Journalism conducted field interviews to supplement a separate analysis of media reporting. They estimated that American drones killed between four hundred and nine hundred and fifty civilians.

The C.I.A.'s position is that these nongovernmental counts are much too high and have been influenced, if inadvertently, by Pakistani government and Taliban propaganda. Early last year, the White House reviewed an internal classified count compiled by the C.I.A. of civilian deaths from drone strikes. Senator Dianne Feinstein, who chairs the Select Committee on Intelligence, disclosed the count's existence at a congressional hearing. She said that classified documents showed that civilian deaths caused by C.I.A. drones each year were "typically in the single digits." The review remains unpublished—in part, a former Administration official said, because the White House couldn't resolve internal debates about the reliability of its methodology. There was also reluctance to publish a specific number, since it would only invite more questions and might expose the scope of classified drone operations. (A C.I.A. spokesman declined to comment for this article.)

The proportion of civilians compared to combatants killed on the ground during American wars since Vietnam has been disputed by researchers. But even the most conservative estimates of civilian casualties place the ratio at one-to-one. In the 1999 NATO-led war in Serbia, where jets used laser-guided and other precision bombs, around five hundred Serbian civilians and three hundred Serbian soldiers were killed, according to the Independent International Commission on Kosovo. The total death toll from drone strikes in Pakistan is estimated at between two thousand and four thousand. Even if one accepts a civilian death toll of nine hundred and fifty-seven—the highest nongovernmental estimate—drones have probably spared more civilians than American jets have in past air wars. And if the numbers Feinstein cited are accurate, drones killed more than twenty fighters for every civilian—a huge leap in precision. Nevertheless, even by that estimate hundreds of families in North and South Waziristan would have suffered the death of an innocent—hardly the foundation for an imagined new age of less provocative American bombing.

There are many reasons to be skeptical of the C.I.A.'s unpublished, lower estimate. According to former Obama Administration officials, the C.I.A.'s Counterterrorism Center, which oversees the agency's drone operations, generates an after-action report, which includes an assessment of whether there was collateral damage. The center has a specialized, independent group that conducts post-strike investigations. The investigators grade the performances of their colleagues and bosses—not exactly a recipe for objectivity. But it seems clear that, over time, the Administration's record improved significantly in avoiding civilian casualties.

In 2008, the last year of the Bush Administration, at least one child was reported killed in a third of all C.I.A. drone strikes in Pakistan, according to the Bureau of Investigative Journalism—a shocking percentage, if it is accurate. In Obama's first year in office, the figure was twenty per cent—still very high. By 2012, it was five per cent.

> "Back in 2007, and in preparation for a deployment to the Middle East, I attended desert warfare school at Creech Air Force Base, Nevada. Along with being an important training facility for Air Force Security Forces personnel, this base also serves as one of the main training and operations centers for the MQ-1 Predator. I will NEVER forget watching those aircraft hover above us during our long and arduous training sessions, and I can certainly attest to the underlying fear factor they bring to the battlefield. Slow and quite deliberate in their movement, they have a particular way of getting into your head."
>
> Kevin Wilson
> Instructor at Rowan Cabarrus Community College
> Master Sergeant, United States Air Force
>
>

Former participants in drone operations attribute the improvement in part to a second-generation armed drone, the Reaper, which allowed operators to scrutinize targets for longer periods on a single flight. Earlier drones carried five-foot-long Hellfire missiles, originally designed to destroy battle tanks; over time, the accuracy of Hellfires improved. The Reaper can also fire a missile, known as the Small Smart Weapon, that is less than two feet long and can take out an individual without killing people in the next room. Strike statistics suggest that, as the C.I.A. improved its procedures for sparing noncombatants, operators fired less often at private homes in North Waziristan, where women and children might also be present. After September, 2010, drones attacked suspected militants riding in vehicles more often than before. "There have been multiple improvements to tailor warheads' blast radiuses to meet particular target characteristics," David Deptula, who served as a deputy chief of staff for intelligence in the Air Force's drone program until 2010, said.

Exactly why and how this change may have been ordered, like much else about the evolution of drone warfare, remains an official secret. The Obama Administration has shielded from public examination essential facts about how often targeting failed and innocents died, and why. In the Bush Administration and the Obama Administration alike, secrecy has defeated public candor and accountability.

Shortly after American Airlines Flight 77 struck the Pentagon, on September 11, 2001, computers at the C.I.A. flashed an instruction: "Immediate Evacuation." From the seventh floor of the Old Headquarters Building, John Rizzo, the agency's highest-ranking career lawyer, watched traffic jams form at the exits from the agency's campus. He decided to stay put, as he recounted in a recent memoir. It was clear that the country had suffered an unprecedented terrorist attack, almost certainly by Al Qaeda, which the C.I.A. had been pursuing for several years. Rizzo took out a yellow legal pad and scribbled notes for what he assumed the agency would now require: a new Memorandum of Notification, or M.O.N., a bedrock document of any C.I.A. covert-action program.

A C.I.A. covert action seeks to surreptitiously influence events abroad while allowing the United States to deny the project's existence. The National Security Act of 1947 legalized covert action as long as a President finds that it does not violate the Constitution or U.S. law and is required to protect the nation. A President must sign a written "Finding" that authorizes the program. Memorandums of Notification are highly classified follow-on documents—specific orders from a President to the C.I.A. describing the scope and the necessity of the operation. They can be concise, sometimes as short as a single page.

On September 17, 2001, President George W. Bush signed a new counterterrorism M.O.N., partly based on Rizzo's input. It was "multiple pages" in length, according to Rizzo. He had worked at the C.I.A. since 1976 and he regarded this document as the "most comprehensive, most ambitious, most aggressive, and most risky Finding or M.O.N. I was ever involved in." Among its provisions, "one short paragraph" authorized targeted killings of Al Qaeda terrorists and their allies. "The language was simple and stark." That paragraph became the foundation for the C.I.A.'s drone operations.

George Tenet, the agency's director at the time, supplemented the M.O.N. with internal guidelines that set down in greater detail how an individual believed to be actively involved in terrorist plots could be nominated and approved for capture or killing. Among other things, the guidelines instructed drone supervisors to avoid civilian casualties "to the maximum extent possible," according to a former senior intelligence official. It was a decidedly lawyerly and elastic standard.

Pakistan's President, Pervez Musharraf, an Army general who had seized power in a 1999 coup, became a trusted partner as the Bush Administration's Global War on Terror unfolded. When American-led forces invaded Afghanistan, in the fall of 2001, many Al Qaeda leaders and foot soldiers—Arabs, Uzbeks, Chechens—escaped into Pakistan. They settled mainly in a region known as FATA, for Federally Administered Tribal Areas, which includes North and South Waziristan and has been a staging ground for jihadist warfare since the nineteen-seventies. Pakistan's generals and politicians, who come mainly from the country's dominant, more developed province of Punjab, treated

Waziristan's residents "as if they were tribes that were living in the Amazon," the journalist Abubakar Siddique, who grew up in the region and is the author of "The Pashtun Question," told me.

In 2002, Musharraf sent Pakistan's Army into South Waziristan to quell Al Qaeda and local sympathizers. In 2004, the Army intensified its operations, and, as violence spread, Musharraf allowed the C.I.A. to fly drones to support Pakistani military action. In exchange, Musharraf told me, the Bush Administration "supplied us helicopters with precision weapons and night-operating capability." He added, "The problem was intelligence collection and targeting. . . . The Americans brought the drones to bear."

Musharraf allowed the C.I.A. to operate drones out of a Pakistani base in Baluchistan. He told me that he often urged Bush Administration officials, "Give the drones to Pakistan." That was not possible, he was told, "because of high-technology transfer restrictions."

On June 17, 2004, a C.I.A. drone killed Nek Muhammad, a Pakistani jihadi leader in Wana, in South Waziristan, who had coöperated with Al Qaeda and led attacks against the Pakistani military. That strike and three subsequent ones in North Waziristan during 2005 were carried out with prior approval from I.S.I., a senior C.I.A. official who served in the region told me. "I would show them the Predator footage and I would say, 'This is what is happening—massive training camps.' " He added, "Every one of these shots was with Pakistani approval."

For several years, the Bush Administration used drone strikes sparingly. At the C.I.A., according to Brian Glyn Williams, a historian at the University of Massachusetts at Dartmouth (his book "Predators: The C.I.A.'s Drone War on Al Qaeda" was published last year), some career officers were reluctant to use drones, because they had "seen the agency burned in the past." Some also feared that too many strikes would destabilize Pakistan and jeopardize Musharraf's position.

In 2006, Bush chose Michael Hayden, a former Air Force general, to be the agency's director. The Counterterrorism Center also got a new leader. His cover name, which has been reported elsewhere, is Roger. Former colleagues describe him as an acerbic, chain-smoking convert to Islam. At a time when Al Qaeda showed signs of revived potency, he argued that drone strikes could weaken its leadership and prevent terrorist attacks on the United States.

That year, an Uzbek militant who had been arrested and detained in Afghanistan told his American interrogators that they should check out the bottom of the wheeled duffel he had been toting when he was taken in. They pulled the bag out of storage and found, according to a senior government official involved in the case, "an eight-and-a-half-by-eleven sheet of paper, folded up," in a hole next to the wheel. It contained Al Qaeda's "code system," the official said—that is, code words used in radio and cell-phone communications to discuss meeting arrangements and plans for violent operations.

The paper also mapped militants' facilities in the vicinity of Miranshah, one of North Waziristan's main towns, the official recalled. The breakthrough led to an ambitious mapping and analysis effort involving satellite photography of Miranshah residences. The National Security Agency searched for the code words in archived transcripts of intercepted phone calls, in order to discover undetected plots and to pinpoint the locations of Al Qaeda leaders. Soon, this effort was also supporting American drone strikes in North Waziristan.

In July, 2008, President Bush approved a plan, proposed by Hayden, to increase drone strikes on Pakistani soil, mainly in North and South Waziristan. Taliban fighters were pouring into Afghanistan from fata, without much interference from Pakistan, to attack American troops. "These sons of bitches are killing Americans. I've had enough," Bush told Hayden, according to Bob Woodward's "Obama's Wars" (2010). No longer would the United States seek permission from Pakistan to strike or notify Pakistani generals in advance. (Musharraf, who had been coddled and protected by Bush for years, was facing impeachment and resigned a month later.)

Hayden approved changes to the internal C.I.A. targeting and strike guidelines. These changes gave rise to what would become known as "signature strikes." The new rules allowed drone operators to fire at armed military-aged males engaged in or associated with suspicious activity even if their identities were unknown. (To justify this looser approach, a former Administration official said, C.I.A. lawyers relied on instructions in an M.O.N. that permitted strikes on terrorist property and facilities.)

Signature strikes are "not a concept known to international humanitarian law," according to ChristofHeyns, the U.N. special rapporteur on extrajudicial, summary, or arbitrary executions. The proper standard for attacking a person under the laws of war is whether the person has a "continuous combat function" or is "directly participating in hostilities." If a signature strike rests on "targeting without sufficient information to make the necessary determination, it is clearly unlawful," Heyns argues in a 2013 report submitted to the U.N. General Assembly. The Obama Administration's position is that, relying on intelligence sources, the C.I.A.'s remote operators could determine whether armed men were involved in violence directed against American personnel and interests. Under the laws of war, which the Administration asserted were applicable in fata, it isn't necessary to know the names of enemy guerrillas before attacking.

After mid-2008, the drone program changed quickly into a more conventional, if unacknowledged, air war. In the three months between August and October, drones struck North and South Waziristan at least twenty times—more strikes than in the previous four years. The C.I.A.'s operators repeatedly hit Al Qaeda or Taliban targets when women and children were present. On September 8, 2008, missiles demolished a North Waziristan home belonging to JalaluddinHaqqani, the defiantly anti-American leader of the Haqqani network. He was absent, but Al Qaeda's chief in Pakistan, Abu Haris, was reportedly killed. Eight women and five children also died, according to a ledger of drone strikes maintained by the FATA government. The ledger records ten other cases of civilian casualties between September and December of 2008. According to the Bureau of Investigative Journalism, children were killed in seven of these instances. When I asked Hayden about those statistics, he replied, "The C.I.A. never had to be reminded of the value of human life. I'm a G.I. and I know the laws of armed conflict: Necessity, distinction, and proportionality are the three rules." Hayden declined to comment specifically about the C.I.A.'s counterterrorism program in

Pakistan, but he added that unmanned aerial vehicles provide "an unblinking stare at a target and the opportunity to be incredibly precise."

In November, Hayden briefed President-elect Obama. According to Gregory Craig, who subsequently joined the new Administration as White House counsel, "Hayden had obviously spent a lot of time, energy, and intellectual resources preparing to explain the drone program—to reassure us about why it was such a successful counterterrorism tool." Hayden's presentation was effective, Craig said. "I walked out of there convinced."

On January 23, 2009, three days after Obama took office, two C.I.A. drones struck inside Pakistan—one in South Waziristan and one in North Waziristan. Both attacks reportedly killed civilians. The strike in North Waziristan hit a private home in the village of Zeraki. According to an affidavit from two witnesses, filed in a complaint to the United Nations Human Rights Council, the dead included an eighth-grade boy and schoolteachers. The South Waziristan strike killed a pro-government peace negotiator who was a tribal leader and four of his family members, entirely in error, according to "Kill or Capture" (2012), a book about Obama's counterterrorism policy by the former Newsweek reporter Daniel Klaidman.

According to Klaidman, John Brennan, at the time Obama's counterterrorism adviser, briefed the President about the South Waziristan mistake and Obama asked how it could have happened, given the weapons' supposed pinpoint accuracy. The President delivered sharp words to C.I.A. officials about the targeting error. Yet he ordered no changes in drone targeting rules at the time, Klaidman reported.

Brennan was a career C.I.A. analyst and a former station chief in Riyadh. He had advised Obama during the 2008 Presidential campaign. Invited to join the Administration, he supported and managed his former agency's advocacy of the targeted killing of suspected terrorists. Brennan also had the experience and the gravitas to push back at the C.I.A. Obama grew to trust him deeply on counterterrorism policy, according to former Administration officials.

To succeed Hayden as the agency's director, Obama selected Leon Panetta, a former chairman of the House Budget Committee, who was known for his bluntness. During the Clinton Administration, he had headed the Office of Management and Budget and served as White House chief of staff. He had barely any intelligence experience. New C.I.A. directors, especially those who are outsiders, are typically advised not to alienate career C.I.A. officers serving on the front lines. Unlike F.B.I. directors, who are appointed to ten-year terms, C.I.A. directors come and go; the agency's bureaucracy has learned to outlast them. Counterterrorism Center leaders saw the drone campaign as their most important operation, and Panetta backed them fully. As he familiarized himself with the C.I.A., Panetta judged the Counterterrorism Center to be "very effective, well run, well resourced, well managed," a former Administration official told me. In 2009, Panetta oversaw some fifty lethal drone attacks; more than half of them produced civilian deaths, according to the Bureau of Investigative Journalism. "American policy was to avoid civilian casualties wherever possible," Panetta wrote in his recently published memoir. An operation that deliberately targeted women or children alongside terrorist suspects "was to be authorized only under extraordinary circumstances." Panetta described cases in which he gave such permission, while seeking to "balance duty to country and respect for life."

In December, 2009, a Jordanian doctor whom the C.I.A. had recruited as an agent inside Al Qaeda blew himself up at an American base in Khost, Afghanistan, killing six agency officers and a contractor. He had really been working for the Pakistani Taliban. The attack was "a big emotional moment for Panetta," the former Administration official said. The following year, the number of drone strikes inside Pakistan doubled. "The C.I.A. really went to war," a participant in White House discussions about the attacks recalled, and Brennan and Obama were supportive. "The White House stood back."

At dusk in Islamabad one evening early this summer, I drove to Fatima Jinnah Park, a vast expanse of footpaths, fields, and eucalyptus trees. It was filled with strolling couples and families on picnics. An acquaintance had arranged a meeting with half a dozen young men from North Waziristan, most of them university students, who had lived through parts of the American drone war in their home villages and towns. In a coffee shop, we sat on plastic chairs in a semicircle and ordered soft drinks. The young men wore polo shirts and bluejeans. Most of them came from relatively privileged tribal families that had suffered during the Taliban's rise to power in Waziristan. The students asked me not to identify them.

Some had come of age as Taliban volunteers. One said that, after 2001, when he was ten or eleven, he carried plastic buckets from house to house to collect money for the organization. Taliban warriors were seen as heroes. "Our parents went to the jihad during the nineteen-eighties," a student said. He meant the C.I.A.-backed campaign against the Soviet occupation of Afghanistan. "Eight or ten years ago, we were all in favor of the Taliban."

That changed. One young man described how the Taliban had kidnapped his father for ransom. Others talked of arbitrary rules and detentions once the Taliban asserted power, after 2007, when they were teen-agers. "We are between two extremes," another student said. "We face regular forces and we also face irregular forces with long hair, beards, and their codes of conduct. It was very difficult to resist them. They imposed their own brand of Islam. If you did not coöperate, you were kidnapped, you were beheaded."

The Taliban are "terrorists," he continued. But he considered the United States a greater menace, because, as the world's leading military power, it "controlled all of this, or could have." But America pursued its own objectives, he said, mainly with drone strikes. "Our economy has been destroyed, our social structure has been destroyed."

"The drones create a lot of misery in our area," one student said. "So do the Arabs." He meant Al Qaeda. "Why are the Arabs coming to our country? Why are they not fighting in their own countries? But we also say to America: If you say the Taliban are terrorists, yes, we agree. They are. But who created them?"

As night fell and we talked on, some of the young men acknowledged that the drone strikes they had seen or heard about from family members have been highly accurate. A few thought that drones offered a better way to bomb Taliban and Al Qaeda leaders in their home towns than F-16s flown by Pakistani pilots, whose bombing could be much more erratic, placing more local civilians at risk. But they also talked about the suffering their families had endured—kidnappings, homes abandoned under pressure—and their own struggles to obtain an education. In their telling, the relative precision of the aircraft that assailed them wasn't the point.

Being attacked by a drone is not the same as being bombed by a jet. With drones, there is typically a much longer prelude to violence. Above North Waziristan, drones circled for hours, or even days, before striking. People below looked up to watch the machines, hovering at about twenty thousand feet, capable of unleashing fire at any moment, like dragon's breath. "Drones may kill relatively few, but they terrify many more," Malik Jalal, a tribal leader in North Waziristan, told me. "They turned the people into psychiatric patients. The F-16s might be less accurate, but they come and go."

Predator and Reaper drones emit what, on the ground, sounds like a flat, gnawing buzz. (Locals sometimes refer to a drone as a bangana, a Pashto word for "wasp.") "In the night, we have seen many times the missile streaking," IhsanDawar, a Pakistani reporter from North Waziristan, told me. "It creates a whoosh-like sound coming out."

The targeted killing of Taliban and Al Qaeda members had a boomerang effect: it spurred the militants to try to identify spies who might have betrayed them. Around North Waziristan's main towns, Miranshah and Mir Ali, which took the brunt of the strikes, paranoia spread.

The Taliban blamed local maliks, government-subsidized tribal leaders who had long presided over the area's war economy—smuggling, arms dealing, mining, and government contracting—often by engaging in corruption. Taliban gunmen seeking control of local rackets executed maliks and their family members in the hundreds. In local bazaars, the Taliban distributed DVDs of their socially superior victims confessing that they had spied for C.I.A. drone operators.

The confessions included elaborate narratives about how the agency supposedly distributed "chips," or homing beacons, to local spies. The spy would toss a chip over a neighbor's wall or into a Taliban jeep, to guide drone missiles to it. The men also confessed that the C.I.A. had given out special pens with invisible ink which were used to mark Taliban vehicles for destruction.

According to the accounts of former detainees, the Taliban tortured their prisoners, so the confessions can hardly be taken at face value. The Taliban also had a powerful motive to force the maliks to admit to spying: such confessions "take the edge off the revenge motivation of the malik's tribe and family," a researcher who grew up in North Waziristan and works in development in Islamabad told me. "People see the video and say, 'Oh, well, if he was a spy tossing around chips, then he deserves to die.' "

Homing beacons are common in policing and espionage. The C.I.A. no doubt uses such devices. Yet it's far from clear whether actual C.I.A. spies in North Waziristan operated by planting chips. The cameras and the telephone-tracking equipment on drones would also allow the C.I.A. to identify and follow targets.

"As a journalist, I haven't seen any chip," Dawar told me. "I don't know if it has any reality behind it or is just a myth." Yet many people believe, he added, that a chip "throws off ultraviolet rays or some kind of magic ray and the missile comes and hits the target."

Not only do many civilians in Waziristan credit the existence of chips but many Taliban do, too. "Once, when I was home, we had a Taliban commander come to our guesthouse and ask to spend the night," the researcher in Islamabad recalled. "He slept in the guesthouse, but he made his driver and two guards sleep in the car and stand around it all night to prevent someone from using the magic pen."

Families in North Waziristan typically live within large walled compounds. Several brothers, their parents, and their extended families might share a single complex. Each compound may contain a hujra, or guesthouse, which usually stands just outside the main wall. In the evening, men gather there to eat dinner and talk war and politics. A rich man signals his status by building a large hujra with comfortable guest rooms for overnight visitors. The less well-heeled might have a hujra with just two rooms, carpets, rope cots, and cushions.

Taliban and Al Qaeda commanders moved from hujra to hujra to avoid detection. The available records of drone strikes make clear that the operators would regularly pick up commanders' movements, follow them to a hujra attached to a private home, watch for hours—or days—and then fire. Many documented strikes took place after midnight, when the target was presumably not moving, children were asleep, and visitors would have returned home.

North Waziristan residents and other Pakistanis I spoke with emphasized how difficult it would be for a drone operator to distinguish between circumstances where a Taliban or Al Qaeda commander had been welcomed into a hujra and where the commander had bullied or forced his way in. If the Taliban "comes to my hujra and asks for shelter, you have no choice," Saleem Safi, a journalist who has travelled extensively in Waziristan, told me. "Now a potential drone target is living in a guest room or a guesthouse on your compound, one wall away from your own house and family."

"You can't protect your family from a strike on a hujra," another resident of North Waziristan said. "Your children will play nearby. They will even go inside to play." The researcher in Islamabad said, "There is always peer pressure, tribal pressure, to be hospitable." He went on, "If you say no, you look like a coward and you lose face. Anyway, you can't say no to them. If a drone strike does take place, you are a criminal in the courts of the Taliban," because you are suspected of espionage and betrayal. "You are also a criminal to the government, because you let the commander sleep in your hujra." In such a landscape, the binary categories recognized by international law—combatant or noncombatant—can seem inadequate to describe the culpability of those who died. Women, children, and the elderly feel pressure from all sides. A young man of military age holding a gun outside a hujra might be a motivated Taliban volunteer, a reluctant conscript, or a victim of violent coercion.

During 2009 and 2010, many of the deaths of children and other civilians recorded contemporaneously by the FATA government occurred during strikes on hujras and homes. Noor Behram's photography and that of other journalists occasionally brought the faces of injured or dead children to public attention, through the Pakistani press and Western human-rights groups. The photos offered a narrative of civilian suffering and became propaganda tools for Taliban media outlets. Targeting errors also became a front in an information war waged by I.S.I. against the United States.

For decades, I.S.I. officers have harbored deep ambivalence about their putative allies at the C.I.A. (According to Pew Research Center opinion polls, a majority of Pakistanis believe that the United States is an enemy of their country.) Beginning in 2009, the Obama Administration, led by the special representative Richard Holbrooke, sought to lessen the mistrust by launching a "strategic dialogue" with Pakistan's military and intelligence leaders, as well as with Pakistan's weak elected civilian politicians. By early 2011, however, that effort had failed. In late January of that year, on a street in Lahore, Raymond Davis, a C.I.A. contractor, shot and killed two men who he believed were trying to kill him, touching off a furor. I.S.I. leaders felt that the C.I.A.'s unilateral operations inside Pakistan had got out of control. Now, when civilians died in drone strikes, I.S.I. helped to whip up public protests.

"That anti-American narrative was basically sponsored by the Army and I.S.I.," Ahmed Rashid, a Pakistani journalist and the author of "Descent Into Chaos" (2008), an investigation of Pakistan's borderlands after the American invasion of Afghanistan, told me. "We all knew it was being orchestrated. The Americans knew, the public knew, the Pakistani media knew. But nobody said anything. Nobody had the courage to say anything." In Washington, evidence that I.S.I. was exploiting C.I.A. strikes to stir anti-American sentiment reduced what incentives the Obama Administration might have had to own up to genuine mistakes: to do so would only play into I.S.I.'s hands.

Still, around this time, as C.I.A. drone operators became more experienced and the technology improved, they shifted away from targeting terrorist "hideouts" and hujras and toward targeting militants' vehicles, according to the FATA ledger of drone strikes. Between March and August, 2010, two-thirds of all C.I.A. drone strikes in North and South Waziristan killed people in compounds or guesthouses. During the next six months, more than half of the strikes hit cars, jeeps, or motorcycles.

The Obama Administration might have benefitted from describing in public how it was adjusting tactics to spare innocent lives. It might have investigated reported errors and compensated survivors, as the U.S. military has done routinely since 2005 whenever it mistakenly kills civilians in Afghanistan. Instead, the law and the logic of secrecy surrounding the C.I.A. campaign silenced the Administration. Jon Stewart riffed freely about drones on "The Daily Show," but at the State Department, a former official there recalled, "we didn't even know if we were allowed to write the word 'drone' in an unclassified e-mail."

The real mistake, according to Ahmed Rashid, was that the C.I.A. went along with Pakistan's hypocrisy in denying that it knew anything about the drone program. To cover up Pakistan's official lies, the United States undermined its own credibility. "Somewhere along the way, the Americans should have drawn a line," Rashid said.

Datta Khel lies about twenty-five miles southwest of Miranshah, toward the border with Afghanistan. Near the town's market and bus depot is an open area suitable for an assembly. On the morning of March 17, 2011, roughly thirty-five maliks, government-approved tribal leaders, had gathered for a jirga, a traditional dispute-resolution meeting. The subject was a feud over a chromite mine.

"There were two tribes in the area, Manzarkhel and Maddakhel," the tribal leader Malik Jalal told me. "The dispute was between these two tribes. They were taking chromite out, but there was a question of who owned what." The Pakistani government knew of the jirga session. Khasadars, or local police, paid by the government, were in attendance, according to court filings.

That morning, Jalal was a little more than two miles away. "I could see the drones in the air, and I actually saw the missiles fly and then heard the explosions," he said. "When I reached the spot, I saw many body parts." The FATA government's contemporaneous ledger of strikes recorded that forty-one people died, and it noted, "The attack was carried out on a jirga and it is feared that all the killed were local tribesmen."

Angry protests erupted in Pakistan. A few Taliban may have been present at the jirga, but the majority were not anti-American fighters, Pakistani officials told reporters at the time. General AshfaqKayani, then Pakistan's Army chief, issued a rare public statement of dissent about C.I.A. operations: "It is highly regrettable that a jirga of peaceful citizens, including elders of the area, was carelessly and callously targeted with complete disregard to human life."

The Obama Administration took a hard line. All of the dead were "terrorists," an anonymous American official told the Times. "These people weren't gathering for a bake sale." The Associated Press quoted an anonymous official offering the same talking point: "This was a group of terrorists, not a charity car wash."

C.I.A. drones had been unusually active in North Waziristan in the days before the jirga strike. On March 11th, drones struck a "suspected vehicle boarded by militants," according to the Pakistani ledger. That evening, a drone bombed a village. Two days later, a drone blew up a "state car" travelling across North Waziristan. On March 16th, a drone destroyed a compound near Datta Khel. From the available evidence, it seems likely that C.I.A. targeting analysts tracked a suspect to the jirga and then decided to kill everyone present.

Afterward, Obama ordered a suspension of C.I.A. drone strikes, according to a former Administration official. No strikes took place in Pakistan for almost a month. But the former official said that Obama authorized an exception to his freeze if the C.I.A. located a "high-value target." When drones struck South Waziristan, in mid-April, a debate erupted in the White House about whether the C.I.A. had violated Obama's order.

Cameron Munter, the U.S. Ambassador in Islamabad at the time, and now a professor at Pomona College, had expressed his deep concern to his intelligence counterparts in both Islamabad and Washington about the extent of drone killing. Munter believed it had got out of hand and was destabilizing Pakistan. According to Mark Mazzetti's "The Way of the Knife" (2013), Munter thought the timing of the jirga strike was "disastrous." After that attack, he argued again to the C.I.A. that it would be a good time to cool down. "The drone strike on March 17th, it exploded—it was just huge," Munter told me. "But, you see, dealing with people at the C.I.A., when I raised it with them, they said, 'You know this is a never-ending war. Whose side are you on?' "

If the Raymond Davis case and the jirga attack strained U.S.-Pakistani relations, the Navy SEAL raid that killed bin Laden, on May 1st, upended them. Kayani and other top generals felt humiliated that they had not been informed in advance. The fact that bin Laden had been hiding less than a mile from Pakistan's principal military academy raised the obvious question of who in Pakistan's establishment might have helped him. In what appeared to be an assertion of the C.I.A.'s freedom to operate independently in Pakistan, the agency resumed drone strikes in North Waziristan five days after the seal raid, at about the same pace as before.

In June, John Brennan appeared at a public seminar on counterterrorism at Johns Hopkins University. Clearly referring to drones, he said, "Nearly for the past year, there hasn't been a single collateral death, because of the exceptional proficiency, precision of the capabilities that we've been able to develop." That claim would have encompassed the jirga strike and a hundred and twenty other strikes in North and South Waziristan dating back to the previous summer, including at least a dozen cases in which nongovernmental researchers have found probable civilian deaths.

On November 16, 2011, C.I.A. drone operators killed a twenty-three-year-old American citizen, Jude Kenan Mohammad, during a nighttime signature strike on a residence in Babar Ghar, in South Waziristan. His death was confirmed only eighteen months later by Attorney General Eric Holder. The C.I.A. apparently did not know of Mohammad's presence in the house.

Mohammad had come of age in Raleigh, North Carolina, and had fallen in with aspiring jihadists; his presence at the site of the attack might indicate that he had become a volunteer soldier. Yet he was not targeted for death as an individual, and, given his American citizenship, any deliberate strike on him would likely have to have been authorized only after an in-depth review overseen by President Obama.

In April, 2012, George Stephanopoulos questioned Brennan on the subject of drones: "Do you stand by the statement you have made in the past that, as effective as they have been, they have not killed a single civilian? That seems hard to believe."

"Well, what I said was that over a period of time before my public remarks that we had no information about a single civilian, a noncombatant, being killed," Brennan replied.

In fact, Brennan had not used the "no information" formula in his remarks at Johns Hopkins the previous year. And his epistemological defense indicated why it has proved impossible to reconcile the large gap between the Administration's count of civilian deaths and those of the Pakistani government and nongovernmental researchers. The C.I.A. has never explained the criteria it uses to count a drone victim as a civilian. Nor has it described what sort of interviews or field research, if any, the agency's analysts undertake to investigate possible mistakes. According to a May, 2012,

Times article by Jo Becker and Scott Shane, "Obama embraced a disputed method for counting civilian casualties that did little to box him in. It in effect counts all military-age males in a strike zone as combatants . . . unless there is explicit intelligence posthumously proving them innocent." In briefings to congressional intelligence committees, the C.I.A. has disputed that characterization, saying that any person deliberately targeted must be associated with a known fighting group or enemy facility, or else be observed preparing for violence.

After the interview with Stephanopoulos, Brennan, during a speech in Washington, further qualified his claims about the precision of drones. "Despite the extraordinary precautions we take, civilians have been accidentally killed," he said. "It is exceedingly rare, but it has happened." He added, "We take it very, very seriously. We go back and we review our actions."

By mid-2012, Obama had ordered Brennan to reassess drone policy. The following spring, at the National Defense University, the President announced a new policy for American drone operations, which remains in effect. The full rulebook is highly classified. Yet Obama did make one new standard public. Before striking, drone operators must determine to a "near certainty" that no civilians are in harm's way—a considerably tougher standard than the C.I.A.'s original one, which dates to the Bush Administration.

Ned Price, a National Security Council spokesman, while declining to discuss any C.I.A. operations, said that Obama's "near certainty" standard was "the highest that we can set." He added, "In those rare instances in which it appears noncombatants may have been killed or injured, after-action reviews have been conducted to determine why, and to insure that we are taking the most effective steps to minimize such risk to noncombatants in the future."

In early 2013, Obama asked Brennan to lead the C.I.A. The President appointed as Brennan's deputy Avril Haines, a National Security Council lawyer who had worked on drone-strike rules and operations. The number of drone strikes carried out in Pakistan fell. Since Brennan became C.I.A. director, according to the data compiled by the Bureau of Investigative Journalism, there has not been a single documented civilian casualty, child or adult, as a result of a drone strike in Waziristan.

Obama's experience of drone war—including the criticism he has received from international lawyers and human-rights groups over civilian casualties—may have motivated him, last year, to tighten targeting oversight, albeit in secret and while evading accountability for errors made early in his Administration. Yet the President and his advisers don't seem to accept how little credit the United States is ever likely to receive from targeted populations just because it chooses to bomb with more accurate drones. On the ground in North Waziristan, drone war doesn't feel much different from other forms of air war, in that many civilians are displaced and frightened, and suffer loss of life and property.

Despite the drone campaign's measurable successes—diminishing the influence of core Al Qaeda, the group around bin Laden that once planned international attacks from Waziristan, and of Al Qaeda in Yemen—the terrorist movement has assumed new shapes in Syria, North Africa, and elsewhere. Pakistan, Yemen, and Somalia are still beset by jihadist violence.

In a research paper published this summer, Micah Zenko and Sarah Kreps, two scholars at the Council on Foreign Relations, argued that the very precision of drone technology raises the prospect for "moral hazard." The reduction in risks may tempt governments to order drones into action more frequently than they would conventional bombers or missiles. In other words, drones may spare more innocents but they may also create more war.

"I think the greatest problem is the mentality that accompanies drone strikes," Philip Alston, an N.Y.U. law professor who investigated drone attacks for the U.N. between 2004 and 2010, told me. "The identification of a list of targets, and if we can succeed in eliminating that list we will have achieved good things—that mentality is what drives it all: if only we can get enough of these bastards, we'll win the war."

One Sunday evening in Islamabad, as pre-monsoon storm clouds blew over the Margalla Hills, I crossed the city's checkpoints to reach the French Club, an oasis notable for the imported liquor in its private bar. I accompanied MirzaShahzad Akbar, a club member. During the Musharraf years, Akbar, a Pakistani lawyer who was trained in London, investigated political corruption for the government. Later, he joined a private firm to serve corporate clients. Four years ago, inspired by an American human-rights lawyer he met in Pakistan, he decided to leave his law firm for the Foundation for Fundamental Rights, which he now leads. He spends his days working with investigators like Noor Behram to collect photographic evidence and sworn testimony about drone-targeting errors, and to advance lawsuits against the Pakistani government, the C.I.A., and sundry American officials.

Akbar is a portly man, partly bald, with a Vandyke beard. Like many members of the Pakistani élite, he is well versed in global media culture. We discussed "Homeland," contenders for the 2016 U.S. Presidential election, and European politics. Later, at his law office, we talked for hours about drones. What sets Akbar apart from Pakistan's privileged class is his passionate advocacy for clients from a population that has virtually no influence or voice even within Pakistan, let alone in Washington: the residents of North and South Waziristan who do not wish to fight in a war.

"When I enter a party, people say, 'It's the Taliban's lawyer,' " he told me. "There are lots of jokes. I've stopped saying what I do." His friends argue vehemently, he continued, that while "it's bad if there are a few civilian casualties" from drones, "there is more damage caused by Pakistani F-16s, and, anyway, if we don't stop them they'll take over and we'll all have beards and our wives and daughters will be in burkas."

When I asked how he answered that argument, he replied, "If we're true liberals, we should also protect the rights of the Taliban." (His foundation does not litigate on behalf of militants or their families.)

Akbar has trouble getting visas to travel to the United States. After he sued the C.I.A., he said, his car was stolen and his office was trashed—events that he assumed were not random. Obama Administration officials, speaking anonymously to the Times, once accused Akbar of fronting for I.S.I. in order to harass the C.I.A. He and his supporters in the West deny that. His foundation's financial support has come mainly from the Bertha Foundation and Reprieve, the British human-rights group. After his lawsuits received widespread publicity in Pakistan, low-level I.S.I. officers visited him a couple of times; he told them that he was serving Pakistan's interests, and they have left him alone.

For Pakistani human-rights advocates, the drone war in Waziristan poses a problem of lesser evils: which is worse, American bombing or Taliban revolution? Taliban suicide bombers have killed thousands of civilians in Pakistan's cities, and the movement is loathed and feared. In June, the Pakistani Army launched a major assault on the Taliban in Waziristan. C.I.A. drones reportedly struck Uzbek militants during the operation. Over all, the Pakistani Army has fought the Taliban to a stalemate, but the group's adherents have gained influence in areas of Karachi, Pakistan's commercial capital, and can still launch successful attacks even in Islamabad. Many Pakistanis understand all too well that their government lacks the competence and the credibility to suppress the Taliban. Some among the élite, therefore, welcome—or, at least, accept—the C.I.A.'s drone strikes as a necessary, temporary compromise.

The private sector use of drones has brought up questions about privacy and safety.

"From Day One, I've been saying, I'm not against drones," Akbar said. "It's just a machine. It's more precise than jets. But it's only as precise as your intelligence." Collecting target information from the sky is difficult; so is gathering information from a semi-hostile partner on the ground, like I.S.I. Akbar wondered aloud if I.S.I., to discredit the United States in the eyes of Pakistanis and the world, might "sometimes give the C.I.A. false targeting information." It would not be surprising.

For as long as the United States does not openly acknowledge targeting errors or pay compensation for victims, and for as long as the Pakistani government lies to the public about its complicity in drone killings, the images of dead civilians that Akbar's investigators collect and publish will resonate. "This is not about taking the Taliban side or the American side," Akbar said. He believes that the United States should hold itself to a higher standard than the Pakistani government. "Our work has been about the fact that there is no transparency or accountability in the U.S. drone program in Pakistan."

Name ______________________ **Course** ______________ **Date** ________________

Chapter Questions

1. From the article, what positives and negatives have been researched about drone warfare?

2. In your opinion, is drone combat more beneficial or problematic to society? Why do you feel this way?

GOOGLE ADDS TO ITS MENAGERIE OF ROBOTS

John Markoff

Boston Dynamics was founded in 1992 by Marc Raibert, a former professor at Massachusetts Institute of Technology. This engineering company designs mobile research robots for the Pentagon and has gained international recognition for their animal-like robots that possess uncanny balance and speed. A few of these machines, the Big Dog and the WildCat, are designed to assist U.S. troops in combat. Extremely agile, some of these robots are able to climb mountains and manage snow covered terrain providing assistance to troops on maneuvers. Google recently acquired Boston Dynamics, adding to its menagerie of autonomous system companies. In the future, these robots might do anything from warehouse work to package delivery or even care for the elderly. In this article, New York Times contributor, John Markoff, explores the endless possibilities these machines offer.

© Ociacia/Shutterstock.com

3D rendering of a futuristic robotic panther

SAN FRANCISCO—BigDog, Cheetah, WildCat and Atlas have joined Google's growing robot menagerie.

Google confirmed on Friday that it had completed the acquisition of Boston Dynamics, an engineering company that has designed mobile research robots for the Pentagon. The company, based in Waltham, Mass., has gained an international reputation for machines that walk with an uncanny sense of balance and even — cheetahlike — run faster than the fastest humans.

It is the eighth robotics company that Google has acquired in the last half-year. Executives at the Internet giant are circumspect about what exactly they plan to do with their robot collection. But Boston Dynamics and its animal kingdom-themed machines bring significant cachet to Google's robotic efforts, which are being led by Andy Rubin, the Google executive who spearheaded the development of Android, the world's most widely used smartphone software.

The deal is also the clearest indication yet that Google is intent on building a new class of autonomous systems that might do anything from warehouse work to package delivery and even elder care.

Boston Dynamics was founded in 1992 by Marc Raibert, a former professor at the Massachusetts Institute of Technology. It has not sold robots commercially, but has pushed the limits of mobile and off-road robotics technology, mostly for Pentagon clients like the Defense Advanced Research Projects Agency, or Darpa. Early on, the company also did consulting work for Sony on consumer robots like the Aibo robotic dog.

Boston Dynamics' walking robots have a reputation for being extraordinarily agile, able to walk over rough terrain and handle surfaces that in some cases are challenging even for humans.

A video of one of its robots named BigDog shows a noisy, gas-powered, four-legged, walking robot that climbs hills, travels through snow, skitters precariously on ice and even manages to stay upright in response to a well-placed human kick. BigDog development started in 2003 in partnership with the British robot maker Foster-Miller, NASA's Jet Propulsion Laboratory and Harvard.

The video has been viewed more than 15 million times since it was posted on YouTube in 2008.

More recently, Boston Dynamics distributed *a video* of a four-legged robot named WildCat, galloping in high-speed circles in a parking lot.

Although the videos frequently inspire comments that the robots will evolve into scary killing machines straight out of the "Terminator" movies, Dr. Raibert has said in the past that he does not consider his company to be a military contractor — it is merely trying to advance robotics technology. Google executives said the company would honor existing military contracts, but that it did not plan to move toward becoming a military contractor on its own.

Under a $10.8 million contract, Boston Dynamics is currently supplying Darpa with a set of humanoid robots named Atlas to participate in the Darpa Robotics Challenge, a two-year contest with a $2 million prize. The contest's goal is creating a class of robots that can operate in natural disasters and catastrophes like the nuclear power plant meltdown in Fukushima, Japan.

"Competitions like the Darpa Robotics Challenge stretch participants to try to solve problems that matter and we hope to learn from the teams' insights around disaster relief," Mr. Rubin said in a statement released by Google.

Boston Dynamics has also designed robots that can climb walls and trees as well as other two- and four-legged walking robots, a neat match to Mr. Rubin's notion that "computers are starting to sprout legs and move around in the environment."

A recent video shows a robot named Cheetah running on a treadmill. This year, the robot was clocked running 29 miles per hour, surpassing the previous legged robot land speed record of 13.1 m.p.h., set in 1999. That's about one mile per hour faster than Jamaica's *Usain Bolt, the two-time Olympic gold medalist* in the 100-meter dash. But it's far short of a real cheetah, which can hit 65 m.p.h.

Google's other robotics acquisitions include companies in the United States and Japan that have pioneered a range of technologies including software for advanced robot arms, grasping technology and computer vision. Mr. Rubin has also said that he is interested in advancing sensor technology.

Mr. Rubin has called his robotics effort a "moonshot," but has declined to describe specific products that might come from the project. He has, however, also said that he does not expect initial product development to go on for years, indicating that Google commercial robots of some nature could be available in the next several years.

Google declined to say how much it paid for its newest robotics acquisition and said that it did not plan to release financial information on any of the other companies it has recently bought.

Dr. Raibert is known as the father of walking robots in the United States. He originally created the Leg Lab, a research laboratory to explore walking machines at Carnegie Mellon University in 1980. He then moved the laboratory to M.I.T. before leaving academia to build engineering systems for the military and Sony.

His research in walking robots began with a pogo-stick project called "the hopper," which he used to test basic concepts.

"I am excited by Andy and Google's ability to think very, very big," Dr. Raibert said, "with the resources to make it happen."

Name ______________________ Course ____________ Date _______________

Chapter Questions

1. How do you feel about robots being used to do "anything from warehouse work to package delivery and even elder care"? What are some pros and cons of this usage?

2. In what other capacities might Google use these animal-like creatures?

THIS SMART RIFLE NEVER MISSES, NOW IN SEMI-AUTOMATIC FORM

Cadie Thompson

One of the most controversial modern weapons is the smart rifle. Built into this new gun's scope is a computer and sensors that allow for precision accuracy and the potential for image recognition. TrackingPoint, the company creating these rifles, claims that this technology, with maximum costs of $27,500 per gun, allows for almost anyone to hit a target up to 1,000 yards away with near 100 percent accuracy. The company is also currently developing a new weapon called the Super Gun that has the potential of hitting a target 1.75 miles away. Cadie Thompson, is a technology reporter on CNBC's Enterprise Team. She discusses how some of these weapons have been bought and are being studied by the United States military and debates the potential impact this technology could have on society.

© Vanoa2/Shutterstock.com

In the near future, you won't even need to know how to fire a rifle to be a crack shot.

At the Consumer Electronic Show, the Austin, Texas-based start-up TrackingPoint showed off its all-new 500 Series AR Smart Rifle, a gun that makes it almost impossible for any user to miss.

TrackingPoint is the inventor of Precision Guided Firearms, a guided shooting system that the company says creates the most accurate guns in the world. The new rifle is the company's first semi-automatic series.

This technology turns even a neophyte into a marksman, at least within a 500-yard range. The user simply "tags" the target, and the gun and ammo do the rest, all for a mere $9,950—the starting price for the new series.

In fact, the system is so accurate that a user will have up to five times the accuracy of an experienced shooter, said Oren Schauble, the company's marketing director.

The gun can track a target moving at up to 10 mph and allows for rapid engagement, meaning a person can shoot multiple targets quickly.

But TrackingPoint doesn't want people to think of it as just a fancy gun maker. The company is really focused on the technology inside the gun, Schauble said.

"Firearms have been the same for decades, the same old design in guns and ammunition," he said.

"What defines our product from the rest of the marketplace is that our firearms integrate technology. We always try to ask ourselves, what in the technology space has not been applied to the arms space and how can we do it?"

TrackingPoint, which rolled out its first Precision Guided Firearm, last year, doesn't just include the technology for a perfect shot. It also integrates other technological functions into it, such as video capabilities and Wi-Fi connectivity. These features allow users to record their hunting expeditions and even share the footage on social media platforms, like Facebook and YouTube.

"The reason we are at CES is because a lot of interest in our technology doesn't come from gun space. Executives and hunters who are in technology make up most of our clients and they see how the technology could be used elsewhere," he said. "Don't be surprised if you see us using this technology for other things."

TrackingPoint makes most of its money from gun sales, but also licenses out some of its technology to Remington. It is also has an assessment contract with the military right now, meaning it's looking to develop technology for them as well.

The company's 500 Series AR Smart Rifle officially launched on Wednesday and is available for pre-order for nearly $10,000. More than 40 orders had been received as of Thursday.

But not everyone who wants the new smart rifle will be able to get their hands on it.

TrackingPoint doesn't sell it's guns to just anybody. Besides the background check that all applicants must go through, the company also has it's own set of standards for purchasers.

The company wants only active users who will provide regular performance feedback on the weapon, Schauble said.

Because the company is still young and the technology is still new, TrackingPoint wants as much user input and sharing of data as possible in order to enhance the weapon's precision and capabilities, he said.

http://www.cnbc.com/2014/01/09/this-smart-rifle-never-misses-now-in-semi-automatic-form.html

Name ____________________ Course ____________ Date ______________

Chapter Questions

1. The smart rifle has the potential for precision accuracy and image recognition so no one except the owner can control the gun. Discuss two advantages and two disadvantages of a weapon with this type of technology, either for the military or for civilians.

2. Overall, what are your feelings about smart weapons? Do you think the addition of the smart rifle in the military is a good idea?

SECTION 5 Science and Medicine

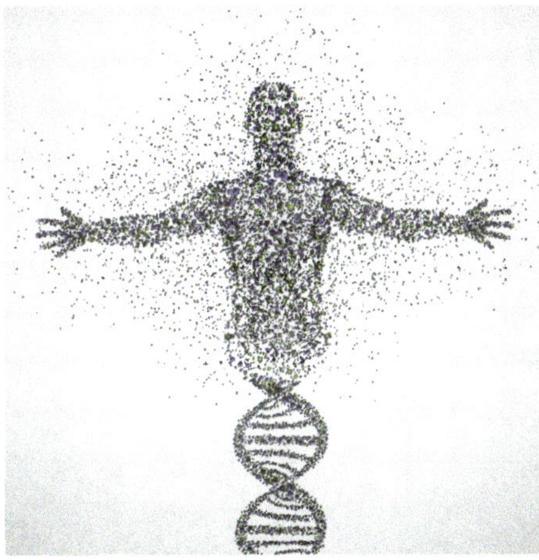

© Lonely/Shutterstock.com

Introduction

Scientific and medical technologies are some of the most controversial yet fastest moving and most innovative technologies of modern culture. There has been much discussion on the positive and negative effects innovations such as cloning, creating "designer" babies, and genetic screening have on society, and the articles and fictional stories in this unit discuss technologies like these.

Breakthroughs in science, particularly genetics, have created ethical and moral consequences that challenge people to decide whether the potential benefits outweigh the possible problems. For example, with advancements in human embryonic stem cell research, the promise is that debilitating diseases could soon be more effectively treated or cured . In 2014, studies showed that over one hundred million Americans suffer from diseases that could, one day, be helped with stem cell therapy. For example, in some cases, injections of human embryonic stem cells have helped restore the sight of blind patients and the injection of stem cells into joints has relieved people of advanced arthritis.

Although there are numerous positive effects of the advancement of these scientific technologies, there are also drawbacks and concerns. For instance, while the use of human embryonic stem cells could be hugely beneficial to some, concerns arise with the use of *human* embryos for such scientific research. Many consider using (and, thus, killing) an embryo for scientific innovations as murder. In this case, people ask the tough question of whether a human embryo should be sacrificed to help others.

When Mary Shelly wrote *Frankenstein* in 1818, the thought of "creating" a human in a laboratory was far-fetched and terrifying, but, with today's technological advancements in science, this, and other innovative ideas, could be a part of the future. The question, in many people's minds, is not whether scientific innovation will happen, but whether it will collectively be a beneficial or potentially harmful conquest. Some of the writers of the readings for this unit seem to look to the future of these technologies with excitement, while others look on with hesitation and, in some instances, even fear.

FRANKENSTEIN, OR THE MODERN PROMETHEUS

Mary Wollstonecraft (Godwin) Shelley

Mary Godwin Shelley was born the illegitimate daughter of William Godwin and feminist Mary Wollstonecraft on August 30, 1797. Her mother died from complications during childbirth, and her father soon remarried. After a turbulent childhood, the sixteen-year-old Mary met and fell in love with the married English poet Percy Bysshe Shelley, and a love affair soon began. Befriending fellow English writer Lord Byron, the young couple spent the summer of 1816 with him and others in Lake Geneva, Switzerland. One rainy, unseasonably cool summer evening, the three writers and other guests began to tell ghost stories. A challenge was given that everyone create a horror tale. As guests begin to weave their stories together, the young Mary struggled; however, later that evening, she dreamt of a creature brought to life with electricity by a scientist with visions of grandeur and aspirations to be godlike. The creature haunted Mary's dream so intensely that, upon waking, she began to write her ghost story from its details, and thus the novel Frankenstein *was born. The fourth chapter of this novel focuses on Dr. Frankenstein's obsession with the creation of life.*

Chapter IV

From this day natural philosophy, and particularly chemistry, in the most comprehensive sense of the term, became nearly my sole occupation. I read with ardour those works, so full of genius and discrimination, which modern inquirers have written on these subjects. I attended the lectures and cultivated the acquaintance of the men of science of the university, and I found even in M. Krempe a great deal of sound sense and real information, combined, it is true, with a repulsive physiognomy and manners, but not on that account the less valuable. In M. Waldman I found a true friend. His gentleness was never tinged by dogmatism, and his instructions were given with an air of frankness and good nature that banished every idea of pedantry. In a thousand ways he smoothed for me the path of knowledge and made the most abstruse inquiries clear and facile to my apprehension. My application was at first fluctuating and uncertain; it gained strength as I proceeded and soon became so ardent and eager that the stars often disappeared in the light of morning whilst I was yet engaged in my laboratory.

As I applied so closely, it may be easily conceived that my progress was rapid. My ardour was indeed the astonishment of the students, and my proficiency that of the masters. Professor Krempe often asked me, with a sly smile, how Cornelius Agrippa went on, whilst M. Waldman expressed the most heartfelt exultation in my progress. Two years passed in this manner, during which I paid no visit to Geneva, but was engaged, heart and soul, in the pursuit of some discoveries which I hoped to make. None but those who have experienced them can conceive of the enticements of science. In other studies you go as far as others have gone before you, and there is nothing more to know; but in a scientific pursuit there is continual food for discovery and wonder. A mind of moderate capacity which closely pursues one study must infallibly arrive at great proficiency in that study; and I, who continually sought the attainment of one object of pursuit and was solely wrapped up in this, improved so rapidly that at the end of two years I made some discoveries in the improvement of some chemical instruments, which procured me great esteem and admiration at the university. When I had arrived at this point and had become as well acquainted with the theory and practice of natural philosophy as depended on the lessons of any of the professors at Ingolstadt, my residence there being no longer conducive to my improvements, I thought of returning to my friends and my native town, when an incident happened that protracted my stay.

One of the phenomena which had peculiarly attracted my attention was the structure of the human frame, and, indeed, any animal endued with life. Whence, I often asked myself, did the principle of life proceed? It was a bold question, and one which has ever been considered as a mystery; yet with how many things are we upon the brink of becoming acquainted, if cowardice or carelessness did not restrain our inquiries. I revolved these circumstances in my mind and determined thenceforth to apply myself more particularly to those branches of natural philosophy which relate to physiology. Unless I had been animated by an almost supernatural enthusiasm, my application to this study would have been irksome and almost intolerable. To examine the causes of life, we must first have recourse to death. I became acquainted with the science of anatomy, but this was not sufficient; I must also observe the natural decay and corruption of the human body. In my education my father had taken the greatest precautions that my mind should be impressed with no supernatural horrors. I do not ever remember to have trembled at a tale of superstition or to have feared the apparition of a spirit. Darkness had no effect upon my fancy, and a churchyard was to me merely the receptacle of bodies deprived of life, which, from being the seat of beauty and strength, had become food for the worm. Now I was led to examine the cause and progress of this decay and forced to spend days and nights in vaults and charnel-houses. My attention was fixed upon every object the most insupportable to the delicacy of the human feelings. I saw how the fine form of man was degraded and wasted; I beheld the corruption of death succeed to the blooming cheek of life; I saw how the worm inherited the wonders of the eye and brain. I paused, examining and analysing all the minutiae of causation, as exemplified in the change from life to death, and death to life, until from the midst of this darkness a sudden light broke in upon me—a light so brilliant and wondrous, yet so simple, that while I became dizzy with the immensity of the prospect which it illustrated, I was surprised that among so many men of genius who had directed their inquiries towards the same science, that I alone should be reserved to discover so astonishing a secret.

Remember, I am not recording the vision of a madman. The sun does not more certainly shine in the heavens than that which I now affirm is true. Some miracle might have produced it, yet the stages of the discovery were distinct and probable. After days and nights of incredible labour and fatigue, I succeeded in discovering the cause of generation and life; nay, more, I became myself capable of bestowing animation upon lifeless matter.

The astonishment which I had at first experienced on this discovery soon gave place to delight and rapture. After so much time spent in painful labour, to arrive at once at the summit of my desires was the most gratifying consummation of my toils. But this discovery was so great and overwhelming that all the steps by which I had been progressively led to it were obliterated, and I beheld only the result. What had been the study and desire of the wisest men since the creation of the world was now within my grasp. Not that, like a magic scene, it all opened upon me at once: the information I had obtained was of a nature rather to direct my endeavours so soon as I should point them towards the object of my search than to exhibit that object already accomplished. I was like the Arabian who had been buried with the dead and found a passage to life, aided only by one glimmering and seemingly ineffectual light.

I see by your eagerness and the wonder and hope which your eyes express, my friend, that you expect to be informed of the secret with which I am acquainted; that cannot be; listen patiently until the end of my story, and you will easily perceive why I am reserved upon that subject. I will not lead you on, unguarded and ardent as I then was, to your destruction and infallible misery. Learn from me, if not by my precepts, at least by my example, how dangerous is the acquirement of knowledge and how much happier that man is who believes his native town to be the world, than he who aspires to become greater than his nature will allow.

When I found so astonishing a power placed within my hands, I hesitated a long time concerning the manner in which I should employ it. Although I possessed the capacity of bestowing animation, yet to prepare a frame for the reception of it, with all its intricacies of fibres, muscles, and veins, still remained a work of inconceivable difficulty and labour. I doubted at first whether I should attempt the creation of a being like myself, or one of simpler organization; but my imagination was too much exalted by my first success to permit me to doubt of my ability to give life to an animal as complex and wonderful as man. The materials at present within my command hardly appeared adequate to so arduous an undertaking, but I doubted not that I should ultimately succeed. I prepared myself for a multitude of reverses; my operations might be incessantly baffled, and at last my work be imperfect, yet when I considered the improvement which every day takes place in science and mechanics, I was encouraged to hope my present attempts would at least lay the foundations of future success. Nor could I consider the magnitude and complexity of my plan as any argument of its impracticability. It was with these feelings that I began the creation of a human being. As the minuteness of the parts formed a great hindrance to my speed, I resolved, contrary to my first intention, to make the being of a gigantic stature, that is to say, about eight feet in height, and proportionably large. After having formed this determination and having spent some months in successfully collecting and arranging my materials, I began.

No one can conceive the variety of feelings which bore me onwards, like a hurricane, in the first enthusiasm of success. Life and death appeared to me ideal bounds, which I should first break through, and pour a torrent of light into our dark world. A new species would bless me as its creator and source; many happy and excellent natures would owe their being to me. No father could claim the gratitude of his child so completely as I should deserve theirs. Pursuing these reflections, I thought that if I could bestow animation upon lifeless matter, I might in process of time (although I now found it impossible) renew life where death had apparently devoted the body to corruption.

© photoneye/Shutterstock.com

These thoughts supported my spirits, while I pursued my undertaking with unremitting ardour. My cheek had grown pale with study, and my person had become emaciated with confinement. Sometimes, on the very brink of certainty, I failed; yet still I clung to the hope which the next day or the next hour might realize. One secret which I alone possessed was the hope to which I had dedicated myself; and the moon gazed on my midnight labours, while, with unrelaxed and breathless eagerness, I pursued nature to her hiding-places. Who shall conceive the horrors of my secret toil as I dabbled among the unhallowed damps of the grave or tortured the living animal to animate the lifeless clay? My limbs now tremble, and my eyes swim with the remembrance; but then a resistless and almost frantic impulse urged me forward; I seemed to have lost all soul or sensation but for this one pursuit. It was indeed but a passing trance, that only made me feel with renewed acuteness so soon as, the unnatural stimulus ceasing to operate, I had returned to my old habits. I collected bones from charnel-houses and disturbed, with profane fingers, the tremendous secrets of the human frame. In a solitary chamber, or rather cell, at the top of the house, and separated from all the other apartments by a gallery and staircase, I kept my workshop of filthy creation; my eyeballs were starting from their sockets in attending to the details of my employment. The dissecting room and the slaughter-house furnished many of my materials; and often did my human nature turn with loathing from my occupation, whilst, still urged on by an eagerness which perpetually increased, I brought my work near to a conclusion.

The summer months passed while I was thus engaged, heart and soul, in one pursuit. It was a most beautiful season; never did the fields bestow a more plentiful harvest or the vines yield a more luxuriant vintage, but my eyes were insensible to the charms of nature. And the same feelings which made me neglect the scenes around me caused me also to forget those friends who were so many miles absent, and whom I had not seen for so long a time. I knew my silence disquieted them, and I well remembered the words of my father: "I know that while you are pleased with yourself you will think of us with affection, and we shall hear regularly from you. You must pardon me if I regard any interruption in your correspondence as a proof that your other duties are equally neglected."

I knew well therefore what would be my father's feelings, but I could not tear my thoughts from my employment, loathsome in itself, but which had taken an irresistible hold of my imagination. I wished, as it were, to procrastinate all that related to my feelings of affection until the great object, which swallowed up every habit of my nature, should be completed.

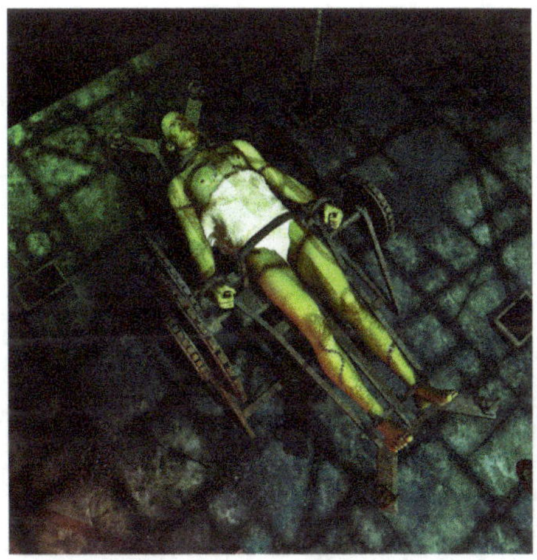
© Bob Orsillo/Shutterstock.com

I then thought that my father would be unjust if he ascribed my neglect to vice or faultiness on my part, but I am now convinced that he was justified in conceiving that I should not be altogether free from blame. A human being in perfection ought always to preserve a calm and peaceful mind and never to allow passion or a transitory desire to disturb his tranquillity. I do not think that the pursuit of knowledge is an exception to this rule. If the study to which you apply yourself has a tendency to weaken your affections and to destroy your taste for those simple pleasures in which no alloy can possibly mix, then that study is certainly unlawful, that is to say, not befitting the human mind. If this rule were always observed; if no man allowed any pursuit whatsoever to interfere with the tranquillity of his domestic affections, Greece had not been enslaved, Caesar would have spared his country, America would have been discovered more gradually, and the empires of Mexico and Peru had not been destroyed.

But I forget that I am moralizing in the most interesting part of my tale, and your looks remind me to proceed. My father made no reproach in his letters and only took notice of my silence by inquiring into my occupations more particularly than before. Winter, spring, and summer passed away during my labours; but I did not watch the blossom or the expanding leaves—sights which before always yielded me supreme delight—so deeply was I engrossed in my occupation. The leaves of that year had withered before my work drew near to a close, and now every day showed me more plainly how well I had succeeded. But my enthusiasm was checked by my anxiety, and I appeared rather like one doomed by slavery to toil in the mines, or any other unwholesome trade than an artist occupied by his favourite employment. Every night I was oppressed by a slow fever, and I became nervous to a most painful degree; the fall of a leaf startled me, and I shunned my fellow creatures as if I had been guilty of a crime. Sometimes I grew alarmed at the wreck I perceived that I had become; the energy of my purpose alone sustained me: my labours would soon end, and I believed that exercise and amusement would then drive away incipient disease; and I promised myself both of these when my creation should be complete.

Name ______________________ Course ______________ Date ________________

Chapter Questions

1. Why do you think Dr. Frankenstein is so obsessed with creating life?

2. Are there situations in society today that are similar to this story? Who could be considered a modern-day Dr. Frankenstein?

THE CASE AGAINST PERFECTION

Michael J. Sandel

Michael J. Sandel is a Professor of Government at Harvard University, and The London Observer *once called him "one of the most popular teachers in the world," because his lectures draw thousands of students eager to discuss modern political issues. His 2004 article, "The Case against Perfection," discusses the topic of designer babies. Sandel presents his readers with questions about this scientific pursuit of perfection and discusses the questionable ethics of genetic enhancements like height, memory, and sex selection. He argues that this form of genetic alteration in a child is socially and biologically flawed. In 2007, Sandel transformed the article into a book entitled* The Case Against Perfection: Ethics in the Age of Genetic Engineering.

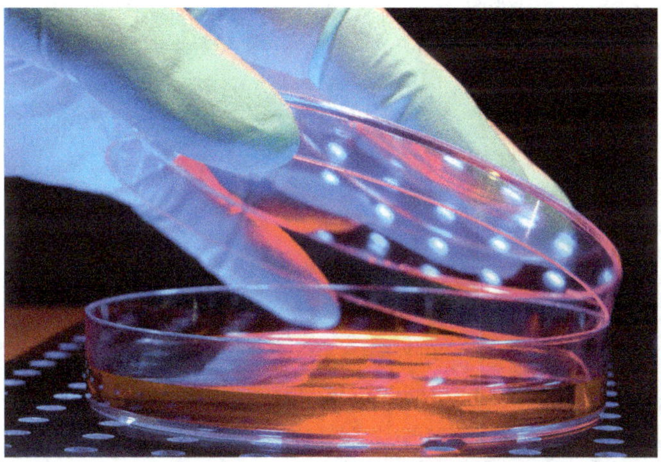

© Olivier Le Queinec/Shutterstock.com

Breakthroughs in genetics present us with a promise and a predicament. The promise is that we may soon be able to treat and prevent a host of debilitating diseases. The predicament is that our new-found genetic knowledge may also enable us to manipulate our own nature—to enhance our muscles, memories, and moods; to choose the sex, height, and other genetic traits of our children; to make ourselves "better than well." When science moves faster than moral understanding, as it does today, men and women struggle to articulate their unease. In liberal societies they reach first for the language of autonomy, fairness, and individual rights. But this part of our moral vocabulary is ill equipped to address the hardest questions posed by genetic engineering. The genomic revolution has induced a kind of moral vertigo.

Consider cloning. The birth of Dolly the cloned sheep, in 1997, brought a torrent of concern about the prospect of cloned human beings. There are good medical reasons to worry. Most scientists agree that cloning is unsafe, likely to produce offspring with serious abnormalities. (Dolly recently died a premature death.) But suppose technology improved to the point where clones were at no greater risk than naturally conceived offspring. Would human cloning still be objectionable? Should our hesitation be moral as well as medical? What, exactly, is wrong with creating a child who is a genetic twin of one parent, or of an older sibling who has tragically died—or, for that matter, of an admired scientist, sports star, or celebrity?

Some say cloning is wrong because it violates the right to autonomy: by choosing a child's genetic makeup in advance, parents deny the child's right to an open future. A similar objection can be raised against any form of bioengineering that allows parents to select or reject genetic characteristics. According to this argument, genetic enhancements for musical talent, say, or athletic prowess, would point children toward particular choices, and so designer children would never be fully free.

At first glance the autonomy argument seems to capture what is troubling about human cloning and other forms of genetic engineering. It is not persuasive, for two reasons. First, it wrongly implies that absent a designing parent, children are free to choose their characteristics for themselves. But none of us chooses his genetic inheritance. The alternative to a cloned or genetically enhanced child is not one whose future is unbound by particular talents but one at the mercy of the genetic lottery.

Second, even if a concern for autonomy explains some of our worries about made-to-order children, it cannot explain our moral hesitation about people who seek genetic remedies or enhancements for themselves. Gene therapy on somatic (that is, non-reproductive) cells, such as muscle cells and brain cells, repairs or replaces defective genes. The moral quandary arises when people use such therapy not to cure a disease but to reach beyond health, to enhance their physical or cognitive capacities, to lift themselves above the norm.

Like cosmetic surgery, genetic enhancement employs medical means for nonmedical ends—ends unrelated to curing or preventing disease or repairing injury, But unlike cosmetic surgery, genetic enhancement is more than skin-deep. If we arc ambivalent about surgery or Botox injections for sagging chins and furrowed brows, we are all the more troubled by genetic engineering for stronger bodies, sharper memories, greater intelligence, and happier moods. The question is whether we are right to be troubled, and if so, on what grounds.

In order to grapple with the ethics of enhancement, we need to confront questions largely lost from view—questions about the moral status of nature, and about the proper stance of human beings toward the given world. Since these questions verge on theology, modern philosophers and political theorists tend to shrink from them. But our new powers of biotechnology make them unavoidable. To see why this is so, consider four examples already on the horizon: muscle enhancement, memory enhancement, growth-hormone treatment, and reproductive technologies that enable parents to choose the sex and some genetic traits of their children. In each case what began as an attempt to treat a disease or prevent a genetic disorder now beckons as an instrument of improvement and consumer choice.

Muscles

Everyone would welcome a gene therapy to alleviate muscular dystrophy and to reverse the debilitating muscle loss that comes with old age. But what if the same therapy were used to improve athletic performance? Researchers have developed a synthetic gene that, when injected into the muscle cells of mice, prevents and even reverses natural muscle deterioration. The gene not only repairs wasted or injured muscles but also strengthens healthy ones. This success bodes well for human applications. H. Lee Sweeney, of the University of Pennsylvania, who leads the research, hopes his discovery will cure the immobility that afflicts the elderly. But Sweeney's bulked-up mice have already attracted the attention of athletes seeking a competitive edge. Although the therapy is not yet approved for human use, the prospect of genetically enhanced weight lifters, home-run sluggers, linebackers, and sprinters is easy to imagine. The widespread use of steroids and other performance-improving drugs in professional sports suggests that many athletes will be eager to avail themselves of genetic enhancement.

Suppose for the sake of argument that muscle-enhancing gene therapy, unlike steroids, turned out to be safe—or at least no riskier than a rigorous weight-training regimen. Would there be a reason to ban its use in sports? There is something unsettling about the image of genetically altered athletes lifting SUVs or hitting 650-foot home runs or running a three-minute mile. But what, exactly, is troubling about it? Is it simply that we find such superhuman spectacles too bizarre to contemplate? Or does our unease point to something of ethical significance?

It might be argued that a genetically enhanced athlete, like a drug-enhanced athlete, would have an unfair advantage over his unenhanced competitors. But the fairness argument against enhancement has a fatal flaw: it has always been the case that some athletes are better endowed genetically than others, and yet we do not consider this to undermine the fairness of competitive sports. From the standpoint of fairness, enhanced genetic differences would be no worse than natural ones, assuming they were safe and made available to all. If genetic enhancement in sports is morally objectionable, it must be for reasons other than fairness.

Memory

Genetic enhancement is possible for brains as well as brawn. In the mid-1990s scientists managed to manipulate a memory-linked gene in fruit flies, creating flies with photographic memories. More recently researchers have produced smart mice by inserting extra copies of a memory-related gene into mouse embryos. The altered mice learn more quickly and remember things longer than normal mice. The extra copies were programmed to remain active even in old age, and the improvement was passed on to offspring. Human memory is more complicated, but biotech companies, including Memory Pharmaceuticals, are in hot pursuit of memory-enhancing drugs, or "cognition enhancers," for human beings. The obvious market for such drugs consists of those who suffer from Alzheimer's and other serious memory disorders. The companies also have their sights on a bigger market: the 81 million Americans over fifty who are beginning to encounter the memory loss that comes naturally with age. A drug that reversed age-related memory loss would be a bonanza for the pharmaceutical industry: a Viagra for the brain. Such use would straddle the line between remedy and enhancement. Unlike a treatment for Alzheimer's, it would cure no disease; but insofar as it restored capacities a person once possessed, it would have a remedial aspect. It could also have purely nonmedical uses: for example, by a lawyer cramming to memorize facts for an upcoming trial, or by a business executive eager to learn Mandarin on the eve of his departure for Shanghai.

Some who worry about the ethics of cognitive enhancement point to the danger of creating two classes of human beings: those with access to enhancement technologies, and those who must make do with their natural capacities. And if the enhancements could be passed down the generations, the two classes might eventually become subspecies—the enhanced and the merely natural. But worry about access ignores the moral status of enhancement itself. Is the scenario troubling because the unenhanced poor would be denied the benefits of bioengineering, or because the enhanced affluent would somehow be dehumanized? As with muscles, so with memory: the fundamental question is not how to ensure equal access to enhancement but whether we should aspire to it in the first place.

Height

Pediatricians already struggle with the ethics of enhancement when confronted by parents who want to make their children taller. Since the 1980s human growth hormone has been approved for children with a hormone deficiency that makes them much shorter than average. But the treatment also increases the height of healthy children. Some parents of healthy children who are unhappy with their stature (typically boys) ask why it should make a difference whether a child is short because of a hormone deficiency or because his parents happen to be short. Whatever the cause, the social consequences are the same.

In the face of this argument some doctors began prescribing hormone treatments for children whose short stature was unrelated to any medical problem. By 1996 such "off-label" use accounted for 40 percent of human-growth-hormone prescriptions. Although it is legal to prescribe drugs for purposes not approved by the Food and Drug Administration, pharmaceutical companies cannot promote such use. Seeking to expand its market, Eli Lilly & Co. recently persuaded the FDA to approve its human growth hormone for healthy children whose projected adult height is in the bottom one percentile—under five feet three inches for boys and four feet eleven inches for girls. This concession raises a large question about the ethics of enhancement: If hormone treatments need not be limited to those with hormone deficiencies, why should they be available only to very short children? Why shouldn't all shorter-than-average children be able to seek treatment? And what about a child of average height who wants to be taller so that he can make the basketball team?

Some oppose height enhancement on the grounds that it is collectively self-defeating; as some become taller, others become shorter relative to the norm. Except in Lake Wobegon, not every child can be above average. As the unenhanced began to feel shorter, they, too, might seek treatment, leading to a hormonal arms race that left everyone worse off, especially those who couldn't afford to buy their way up from shortness.

But the arms-race objection is not decisive on its own. Like the fairness objection to bioengineered muscles and memory, it leaves unexamined the attitudes and dispositions that prompt the drive for enhancement. If we were bothered only by the injustice of adding shortness to the problems of the poor, we could remedy that unfairness by publicly subsidizing height enhancements. As for the relative height deprivation suffered by innocent bystanders, we could compensate them by taxing those who buy their way to greater height. The real question is whether we want to live in a society where parents feel compelled to spend a fortune to make perfectly healthy kids a few inches taller.

Sex Selection

Perhaps the most inevitable nonmedical use of bioengineering is sex selection. For centuries parents have been trying to choose the sex of their children. Today biotech succeeds where folk remedies failed.

One technique for sex selection arose with prenatal tests using amniocentesis and ultrasound. These medical technologies were developed to detect genetic abnormalities such as spina bifida and Down syndrome. But they can also reveal the sex of the fetus—allowing for the abortion of a fetus of an undesired sex. Even among those who favor abortion rights, few advocate abortion simply because the parents do not want a girl. Nevertheless, in traditional societies with a powerful cultural preference for boys, this practice has become widespread.

Sex selection need not involve abortion, however. For couples undergoing *in vitro* fertilization (IVF), it is possible to choose the sex of the child before the fertilized egg is implanted in the womb. One method makes use of preimplantation genetic diagnosis (PGD), a procedure developed to screen for genetic diseases. Several eggs are fertilized in a petri dish and grown to the eight-cell stage (about three days). At that point the embryos are tested to determine their sex. Those of the desired sex are implanted: the others are typically discarded. Although few couples are likely to undergo the difficulty and expense of IVF simply to choose the sex of their child, embryo screening is a highly reliable means of sex selection. And as our genetic knowledge increases, it may be possible to use PGD to cull embryos carrying undesired genes, such as those associated with obesity, height, and skin color. The sciencefiction movie *Gattaca* depicts a future in which parents routinely screen embryos for sex, height, immunity to disease, and even IQ. There is something troubling about the *Gattaca* scenario, but it is not easy to identify what exactly is wrong with screening embryos to choose the sex of our children.

One line of objection draws on arguments familiar from the abortion debate. Those who believe that an embryo is a person reject embryo screening for the same reasons they reject abortion. If an eight-cell embryo growing in a petri dish is morally equivalent to a fully developed human being, then discarding it is no better than aborting a fetus, and both practices are equivalent to infanticide. Whatever its merits, however, this "pro-life" objection is not an argument against sex selection as such.

The latest technology poses the question of sex selection unclouded by the matter of an embryo's moral status. The Genetics & IVF Institute, a for-profit infertility clinic in Fairfax, Virginia, now offers a sperm-sorting technique that makes it possible to choose the sex of one's child before it is conceived. X-bearing sperm, which produce girls, carry more DNA than Y-bearing sperm, which produce boys; a device called a flow cytometer can separate them. The process, called MicroSort, has a high rate of success.

If sex selection by sperm sorting is objectionable, it must be for reasons that go beyond the debate about the moral status of the embryo. One such reason is that sex selection is an instrument of sex discrimination—typically against girls, as illustrated by the chilling sex ratios in India and China. Some speculate that societies with substantially more men than women will be less stable, more violent, and more prone to crime or war. These are legitimate worries—but the sperm-sorting company has a clever way of addressing them. It offers MicroSort only to couples who want to choose the sex of a child for purposes of "family balancing." Those with more sons than daughters may choose a girl, and vice versa. But customers may not use the technology to stock up on children of the same sex, or even to choose the sex of their firstborn child. (So far the majority of MicroSort clients have chosen girls.) Under restrictions of this kind, do any ethical issues remain that should give us pause?

The case of MicroSort helps us isolate the moral objections that would persist if muscle-enhancement, memory-enhancement, and height-enhancement technologies were safe and available to all.

It is commonly said that genetic enhancements undermine our humanity by threatening our capacity to act freely, to succeed by our own efforts, and to consider ourselves responsible—worthy of praise or blame—for the things we do and for the way we are. It is one thing to hit seventy home runs as the result of disciplined training and effort, and something else, something less, to hit them with the help of steroids or genetically enhanced muscles. Of course, the roles of effort and enhancement will be a matter of degree. But as the role of enhancement increases, our admiration for the achievement fades—or, rather, our admiration for the achievements shifts from the player to his pharmacist. This suggests that our moral response to enhancement is a response to the diminished agency of the person whose achievement is enhanced.

Though there is much to be said for this argument, I do not think the main problem with enhancement and genetic engineering is that they undermine effort and erode human agency. The deeper danger is that they represent a kind of hyperagency—a Promethean aspiration to remake nature, including human nature, to serve our purposes and satisfy our desires. The problem is not the drift to mechanism but the drive to mastery. And what the drive to mastery misses and may even destroy is an appreciation of the gifted character of human powers and achievements.

To acknowledge the giftedness of life is to recognize that our talents and powers are not wholly our own doing, despite the effort we expend to develop and to exercise them. It is also to recognize that not everything in the world is open to whatever use we may desire or devise. Appreciating the gifted quality of life constrains the Promethean project and conduces to a certain humility. It is in part a religious sensibility. But its resonance reaches beyond religion.

It is difficult to account for what we admire about human activity and achievement without drawing upon some version of this idea. Consider two types of athletic achievement. We appreciate players like Pete Rose, who are not blessed with great natural gifts but who manage, through striving, grit, and determination, to excel in their sport. But we also admire players like Joe DiMaggio, who display natural gifts with grace and effortlessness. Now, suppose we learned that both players took performance-enhancing drugs. Whose turn to drugs would we find more deeply disillusioning? Which aspect of the athletic ideal—effort or gift—would be more deeply offended?

Some might say effort: the problem with drugs is that they provide a shortcut, a way to win without striving. But striving is not the point of sports; excellence is. And excellence consists at least partly in the display of natural talents and gifts that are no doing of the athlete who possesses them. This is an uncomfortable fact for democratic societies. We want to believe that success, in sports and in life, is something we earn, not something we inherit. Natural gifts, and the admiration they inspire, embarrass the meritocratic faith; they cast doubt on the conviction that praise and rewards flow from effort alone. In the face of this embarrassment we inflate the moral significance of striving, and depreciate giftedness. This distortion can be seen, for example, in network-television coverage of the Olympics, which focuses less on the feats the athletes perform than on heartrending stories of the hardships they have overcome and the struggles they have waged to triumph over an injury or a difficult upbringing or political turmoil in their native land.

But effort isn't everything. No one believes that a mediocre basketball player who works and trains even harder than Michael Jordan deserves greater acclaim or a bigger contract. The real problem with genetically altered athletes is that they corrupt athletic competition as a human activity that honors the cultivation and display of natural talents. From this standpoint, enhancement can be seen as the ultimate expression of the ethic of effort and willfulness—a kind of high-tech striving. The ethic of willfulness and the biotechnological powers it now enlists are arrayed against the claims of giftedness.

The ethic of giftedness, under siege in sports, persists in the practice of parenting. But here, too, bioengineering and genetic enhancement threaten to dislodge it. To appreciate children as gifts is to accept them as they come, not as objects of our design or products of our will or instruments of our ambition. Parental love is not contingent on the talents and attributes a child happens to have. We choose our friends and spouses at least partly on the basis of qualities we find attractive. But we do not choose our children. Their qualities are unpredictable, and even the most conscientious parents cannot be held wholly responsible for the kind of children they have. That is why parenthood, more than other human relationships, teaches what the theologian William F. May calls an "openness to the unbidden."

May's resonant phrase helps us see that the deepest moral objection to enhancement lies less in the perfection it seeks than in the human disposition it expresses and promotes. The problem is not that parents usurp the autonomy of a child they design. The problem lies in the hubris of the designing parents, in their drive to master the mystery of birth. Even if this disposition did not make parents tyrants to their children, it would disfigure the relation between parent and child, and deprive the parent of the humility and enlarged human sympathies that an openness to the unbidden can cultivate.

To appreciate children as gifts or blessings is not, of course, to be passive in the face of illness or disease. Medical intervention to cure or prevent illness or restore the injured to health does not desecrate nature but honors it. Healing sickness or injury does not override a child's natural capacities but permits them to flourish.

Nor does the sense of life as a gift mean that parents must shrink from shaping and directing the development of their child. Just as athletes and artists have an obligation to cultivate their talents, so parents have an obligation to cultivate their children, to help them discover and develop their talents and gifts. As May points out, parents give their children two kinds of love: accepting love and transforming love. Accepting love affirms the being of the child, whereas transforming love seeks the well-being of the child. Each aspect corrects the excesses of the other, he writes: "Attachment becomes too quietistic if it slackens into mere acceptance of the child as he is." Parents have a duty to promote their children's excellence.

These days, however, overly ambitious parents are prone to get carried away with transforming love—promoting and demanding all manner of accomplishments from their children, seeking perfection. "Parents find it difficult to maintain an equilibrium between the two sides of love," May observes. "Accepting love, without transforming love, slides into indulgence and finally neglect. Transforming love, without accepting love, badgers and finally rejects." May finds in these competing impulses a parallel with modern science: it, too, engages us in beholding the given world, studying and savoring it, and also in molding the world, transforming and perfecting it.

The mandate to mold our children, to cultivate and improve them, complicates the case against enhancement. We usually admire parents who seek the best for their children, who spare no effort to help them achieve happiness and success. Some parents confer advantages on their children by enrolling them in expensive schools, hiring private tutors, sending them to tennis camp, providing them with piano lessons, ballet lessons, swimming lessons, SAT-prep courses, and so on. If it is permissible and even admirable for parents to help their children in these ways, why isn't it equally admirable for parents to use whatever genetic technologies may emerge (provided they are safe) to enhance their children's intelligence, musical ability, or athletic prowess?

The defenders of enhancement are right to this extent: improving children through genetic engineering is similar in spirit to the heavily managed, high-pressure child-rearing that is now common. But this similarity does not vindicate genetic enhancement. On the contrary, it highlights a problem with the trend toward hyperparenting. One conspicuous example of this trend is sports-crazed parents bent on making champions of their children. Another is the frenzied drive of over-bearing parents to mold and manage their children's academic careers.

As the pressure for performance increases, so does the need to help distractible children concentrate on the task at hand. This may be why diagnoses of attention deficit and hyperactivity disorder have increased so sharply. Lawrence Diller, a pediatrician and the author of *Running on Ritalin*, estimates that five to six percent of American children under eighteen (a total of four to five million kids) are currently prescribed Ritalin, Adderall, and other stimulants, the treatment of choice for ADHD. (Stimulants counteract hyperactivity by making it easier to focus and sustain attention.) The number of Ritalin prescriptions for children and adolescents has tripled over the past decade, but not all users suffer from attention disorders or hyperactivity. High school and college students have learned that prescription stimulants improve concentration for those with normal attention spans, and some buy or borrow their classmates' drugs to enhance their performance on the SAT or other exams. Since stimulants work for both medical and non-medical purposes, they raise the same moral questions posed by other technologies of enhancement.

However those questions are resolved, the debate reveals the cultural distance we have traveled since the debate over marijuana, LSD, and other drugs a generation ago. Unlike the drugs of the 1960s and 1970s Ritalin and Adderall are not for checking out but for buckling down, not for beholding the world and taking it in but for molding the world and fitting in. We used to speak of nonmedical drug use as "recreational." That term no longer applies. The steroids and stimulants that figure in the enhancement debate are not a source of recreation but a bid for compliance—a way of answering a competitive society's demand to improve our performance and perfect our nature. This demand for performance and perfection animates the impulse to rail against the given. It is the deepest source of the moral trouble with enhancement.

Some see a clear line between genetic enhancement and other ways that people seek improvement in their children and themselves. Genetic manipulation seems somehow worse—more intrusive, more sinister—than other ways of enhancing performance and seeking success. But morally speaking, the difference is less significant than it seems. Bioengineering gives us reason to question the low-tech, high-pressure child-rearing practices we commonly accept. The hyperparenting familiar in our time represents an anxious excess of mastery and dominion that misses the sense of life as a gift. This draws it disturbingly close to eugenics.

The shadow of eugenics hangs over today's debates about genetic engineering and enhancement. Critics of genetic engineering argue that human cloning, enhancement, and the quest for designer children are nothing more than "privatized" or "free-market" eugenics. Defenders of enhancement reply that genetic choices freely made are not really eugenic—at least not in the pejorative sense. To remove the coercion, they argue, is to remove the very thing that makes eugenic policies repugnant.

Sorting out the lesson of eugenics is another way of wrestling with the ethics of enhancement. The Nazis gave eugenics a bad name. But what, precisely, was wrong with it? Was the old eugenics objectionable only insofar as it was coercive? Or is there something inherently wrong with the resolve to deliberately design our progeny's traits?

James Watson, the biologist who, with Francis Crick, discovered the structure of DNA, sees nothing wrong with genetic engineering and enhancement, provided they are freely chosen rather than state-imposed. And yet Watson's language contains more than a whiff of the old eugenic sensibility. "If you really are stupid. I would call that a disease," he recently told *The Times of London*. "The lower 10 percent who really have difficulty, even in elementary school, what's the cause of it? A lot of people would like to say, "Well, poverty, things like that." It probably isn't. So I'd like to get rid of that, to help the lower 10 percent." A few years ago Watson stirred controversy by saying that if a gene for homosexuality were discovered, a woman should be free to abort a fetus that carried it. When his remark provoked an uproar, he replied that he was not singling out gays but asserting a principle: women should be free to abort fetuses for any reason of genetic preference—for example, if the child would be dyslexic, or lacking musical talent, or too short to play basketball.

Watson's scenarios are clearly objectionable to those for whom all abortion is an unspeakable crime. But for those who do not subscribe to the pro-life position, these scenarios raise a hard question: If it is morally troubling to contemplate abortion to avoid a gay child or a dyslexic one, doesn't this suggest that something is wrong with acting on any eugenic preference, even when no state coercion is involved?

Consider the market in eggs and sperm. The advent of artificial insemination allows prospective parents to shop for gametes with the genetic traits they desire in their off-spring. It is a less predictable way to design children than cloning or pre-implantation genetic screening, but it offers a good example of a procreative practice in which the old eugenics meets the new consumerism. A few years ago some Ivy League newspapers ran an ad seeking an egg from a woman who was at least five feet ten inches tall and athletic, had no major family medical problems, and had a combined SAT score of 1400 or above. The ad offered $50,000 for an egg from a donor with these traits. More recently a Web site was launched claiming to auction eggs from fashion models whose photos appeared on the site, at starting bids of $15,000 to $150,000.

On what grounds, if any, is the egg market morally objectionable? Since no one is forced to buy or sell, it cannot be wrong for reasons of coercion. Some might worry that hefty prices would exploit poor women by presenting them with an offer they couldn't refuse. But the designer eggs that fetch the highest prices are likely to be sought from the privileged, not the poor. If the market for premium eggs gives us moral qualms, this, too, shows that concerns about eugenics arc not put to rest by freedom of choice.

A tale of two sperm banks helps explain why. The Repository for Germinal Choice, one of America's first sperm banks, was not a commercial enterprise. It was opened in 1980 by Robert Graham, a philanthropist dedicated to improving the world's "germ plasm" and counteracting the rise of "retrograde humans." His plan was to collect the sperm of Nobel Prize-winning scientists and make it available to women of high intelligence, in hopes of breeding super-smart babies. But Graham had trouble persuading Nobel laureates to donate their sperm for his bizarre scheme, and so settled for sperm from young scientists of high promise. His sperm bank closed in 1999.

In contrast, California Cryobank, one of the world's leading sperm banks, is a for-profit company with no overt eugenic mission. Cappy Rothman, M.D., a co-founder of the firm, has nothing but disdain for Graham's eugenics, although the standards Cryobank imposes on the sperm it recruits are exacting. Cryobank has offices in Cambridge, Massachusetts, between Harvard and MIT, and in Palo Alto, California, near Stanford. It advertises for donors in campus newspapers (compensation up to $900 a month), and accepts less than five percent of the men who apply. Cryo-bank's marketing materials play up the prestigious source of its sperm. Its catalogue provides detailed information about the physical characteristics of each donor, along with his ethnic origin and college major. For an extra fee prospective customers can buy the results of a test that assesses the donor's temperament and character type. Rothman reports that Cryobank's ideal sperm donor is six feet tall, with brown eyes, blond hair, and dimples, and has a college degree—not because the company wants to propagate those traits, but because those are the traits his customers want: "If our customers wanted high school dropouts, we would give them high school dropouts."

Not everyone objects to marketing sperm. But anyone who is troubled by the eugenic aspect of the Nobel Prize sperm bank should be equally troubled by Cryobank, consumer-driven though it be. What, after all, is the moral difference between designing children according to an explicit eugenic purpose and designing children according to the dictates of the market? Whether the aim is to improve humanity's "germ plasm" or to cater to consumer preferences, both practices are eugenic insofar as both make children into products of deliberate design.

A number of political philosophers call for a new "liberal eugenics." They argue that a moral distinction can be drawn between the old eugenic policies and genetic enhancements that do not restrict the autonomy of the child. "While old-fashioned authoritarian eugenicists sought to produce citizens out of a single centrally designed mould," writes Nicholas Agar, "the distinguishing mark of the new liberal eugenics is state neutrality." Government may not tell parents what sort of children to design, and parents may engineer in their children only those traits that improve their capacities without biasing their choice of life plans. A recent text on genetics and justice, written by the bioethicists Allen Buchanan, Dan W. Brock, Norman Daniels, and Daniel Wilder, offers a similar view. The "bad reputation of eugenics," they write, is due to practices that "might be avoidable in a future eugenic program." The problem with the old eugenics was that its burdens fell disproportionately on the weak and the poor, who were unjustly sterilized and segregated. But provided that the benefits and burdens of genetic improvement are fairly distributed, these bioethicists argue, eugenic measures are unobjectionable and may even be morally required.

The libertarian philosopher Robert Nozick proposed a "genetic supermarket" that would enable parents to order children by design without imposing a single design on the society as a whole: "This supermarket system has the great virtue that it involves no centralized decision fixing the future human type(s)."

Even the leading philosopher of American liberalism, John Rawls, in his classic *A Theory of Justice* (1971), offered a brief endorsement of noncoercive eugenics. Even in a society that agrees to share the benefits and burdens of the genetic lottery, it is "in the interest of each to have greater natural assets," Rawls wrote. "This enables him to pursue a preferred plan of life." The parties to the social contract "want to insure for their descendants the best genetic endowment (assuming their own to be fixed)." Eugenic policies are therefore not only permissible but required as a matter of justice. "Thus over time a society is to take steps at least to preserve the general level of natural abilities and to prevent the diffusion of serious defects."

But removing the coercion does not vindicate eugenics. The problem with eugenics and genetic engineering is that they represent the one-sided triumph of willfulness over giftedness, of dominion over reverence, of molding over beholding. Why, we may wonder, should we worry about this triumph? Why not shake off our unease about genetic enhancement as so much superstition? What would be lost if biotechnology dissolved our sense of giftedness?

From a religious standpoint the answer is clear. To believe that our talents and powers are wholly our own doing is to misunderstand our place in creation, to confuse our role with God's. Religion is not the only source of reasons to care about giftedness, however. The moral stakes can also be described in secular terms. If bioengineering made the myth of the "self-made man" come true, it would be difficult to view our talents as gifts for which we are indebted, rather than as achievements for which we are responsible. This would transform three key features of our moral landscape: humility, responsibility, and solidarity.

In a social world that prizes mastery and control, parenthood is a school for humility. That we care deeply about our children and yet cannot choose the kind we want teaches parents to be open to the unbidden. Such openness is a disposition worth affirming, not only within families but in the wider world as well. It invites us to abide the unexpected, to live with dissonance, to rein in the impulse to control. A *Gattaca*-like world in which parents became accustomed to specifying the sex and genetic traits of their children would be a world inhospitable to the unbidden, a gated community writ large. The awareness that our talents and abilities are not wholly our own doing restrains our tendency toward hubris.

Though some maintain that genetic enhancement erodes human agency by overriding effort, the real problem is the explosion, not the erosion, of responsibility. As humility gives way, responsibility expands to daunting proportions. We attribute less to chance and more to choice. Parents become responsible for choosing, or failing to choose, the right traits for their children. Athletes become responsible for acquiring, or failing to acquire, the talents that will help their teams win.

One of the blessings of seeing ourselves as creatures of nature, God, or fortune is that we are not wholly responsible for the way we are. The more we become masters of our genetic endowments, the greater the burden we bear for the talents we have and the way we perform. Today when a basketball player misses a rebound, his coach can blame him for being out of position. Tomorrow the coach may blame him for being too short, Even now the use of performance-enhancing drugs in professional sports is subtly transforming the expectations players have for one another; on some teams players who take the field free from amphetamines or other stimulants are criticized for "playing naked."

The more alive we are to the chanced nature of our lot, the more reason we have to share our fate with others. Consider insurance. Since people do not know whether or when various ills will befall them, they pool their risk by buying health insurance and life insurance. As life plays itself out, the healthy wind up subsidizing the unhealthy, and those who live to a ripe old age wind up subsidizing the families of those who die before their time. Even without a sense of mutual obligation, people pool their risks and resources and share one another's fate.

But insurance markets mimic solidarity only insofar as people do not know or control their own risk factors. Suppose genetic testing advanced to the point where it could reliably predict each person's medical future and life expectancy. Those confident of good health and long life would opt out of the pool, causing other people's premiums to skyrocket. The solidarity of insurance would disappear as those with good genes fled the actuarial company of those with bad ones.

The fear that insurance companies would use genetic data to assess risks and set premiums recently led the Senate to vote to prohibit genetic discrimination in health insurance. But the bigger danger, admittedly more speculative, is that genetic enhancement, if routinely practiced, would make it harder to foster the moral sentiments that social solidarity requires.

Why, after all, do the successful owe anything to the least-advantaged members of society? The best answer to this question leans heavily on the notion of giftedness. The natural talents that enable the successful to flourish are not their own doing but, rather, their good fortune—a result of the genetic lottery. If our genetic endowments are gifts, rather than achievements for which we can claim credit, it is a mistake and a conceit to assume that we are entitled to the full measure of the bounty they reap in a market economy. We therefore have an obligation to share this bounty with those who, through no fault of their own, lack comparable gifts.

A lively sense of the contingency of our gifts—a consciousness that none of us is wholly responsible for his or her success—saves a meritocratic society from sliding into the smug assumption that the rich are rich because they are more deserving than the poor. Without this, the successful would become even more likely than they are now to view themselves as self-made and self-sufficient, and hence wholly responsible for their success. Those at the bottom of society would be viewed not as disadvantaged, and thus worthy of a measure of compensation, but as simply unfit, and thus worthy of eugenic repair. The meritocracy, less chastened by chance, would become harder, less forgiving. As perfect genetic knowledge would end the simulacrum of solidarity in insurance markets, so perfect genetic control would erode the actual solidarity that arises when men and women reflect on the contingency of their talents and fortunes.

Thirty-five years ago Robert L. Sinsheimer, a molecular biologist at the California Institute of Technology, glimpsed the shape of things to come. In an article titled "The Prospect of Designed Genetic Change" he argued that freedom of choice would vindicate the new genetics, and set it apart from the discredited eugenics of old.

> To implement the older eugenics would have required a massive social programme carried out over many generations. Such a programme could not have been initiated without the consent and cooperation of a major fraction of the population, and would have been continuously subject to social control. In contrast, the new eugenics could, at least in principle, he implemented on a quite individual basis, in one generation, and subject to no existing restrictions.

According to Sinsheimer, the new eugenics would be voluntary rather than coerced, and also more humane. Rather than segregating and eliminating the unfit, it would improve them. "The old eugenics would have required a continual selection for breeding of the fit, and a culling of the unfit," he wrote. "The new eugenics would permit in principle the conversion of all the unfit to the highest genetic level."

Sinsheimer's paean to genetic engineering caught the heady, Promethean self-image of the age. He wrote hopefully of rescuing "the losers in that chromosomal lottery that so firmly channels our human destinies," including not only those born with genetic defects but also "the 50,000,000 'normal' Americans with an IQ of less than 90." But he also saw that something bigger than improving on nature's "mindless, age-old throw of dice" was at stake. Implicit in technologies of genetic intervention was a more exalted place for human beings in the cosmos. "As we enlarge man's freedom, we diminish his constraints and that which he must accept as given," he wrote, Copernicus and Darwin had "demoted man from his bright glory at the focal point of the universe," but the new biology would restore his central role. In the mirror of our genetic knowledge we would see ourselves as more than a link in the chain of evolution: "We can be the agent of transition to a whole new pitch of evolution. This is a cosmic event."

There is something appealing, even intoxicating, about a vision of human freedom unfettered by the given. It may even be the case that the allure of that vision played a part in summoning the genomic age into being. It is often assumed that the powers of enhancement we now possess arose as an inadvertent by-product of biomedical progress—the genetic revolution came, so to speak, to cure disease, and stayed to tempt us with the prospect of enhancing our performance, designing our children, and perfecting our nature. That may have the story backwards. It is more plausible to view genetic engineering as the ultimate expression of our resolve to see ourselves astride the world, the masters of our nature. But that promise of mastery is flawed. It threatens to banish our appreciation of life as a gift, and to leave us with nothing to affirm or behold outside our own will.

Name ______________________ Course ____________ Date ______________

Chapter Questions

1. The author presents both negative and positive aspects of designer babies. Do you think designer babies should be unregulated and allowed? Explain.

2. Would you ever consider "designing" a baby for yourself? If so, what characteristics would you choose for your child and why?

3. Why do you think some people have a moral problem with the scientific and genetic technologies mentioned in this reading?

THE EXACT LOCATION OF THE SOUL

Richard Selzer

Richard Selzer is a physician and writer who explores what it means to "heal" rather than just medicate patients. In this writing he transforms his experience as a surgeon into literary meditations.

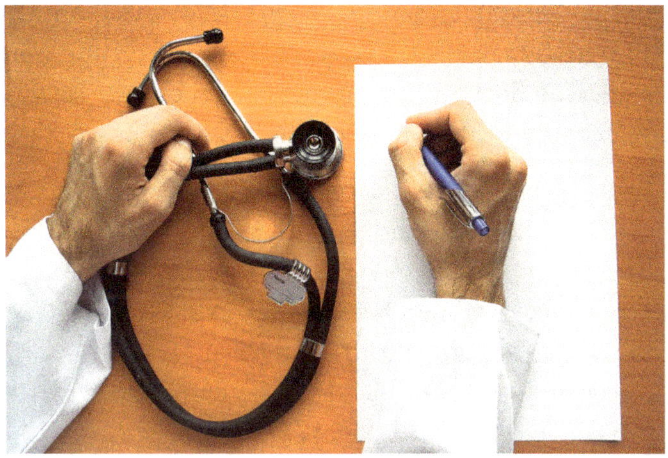

© Kostiantyn Karpenko/Shutterstock.com

The Exact Location of the Soul

SOMEONE asked me why a surgeon would write. Why, when the shelves are already too full? They sag under the deadweight of books. To add a single adverb is to risk exceeding the strength of the boards. A surgeon should abstain. A surgeon, whose fingers are more at home in the steamy gulies of the body than they are tapping the dry keys of a typewriter. A surgeon, who feels the slow slide of intestines against the back of his hand and is no more alarmed than were a family of snakes taking their comfort from such an adolent rubbing. A surgeon, who palms the human heart as though it were some captured bird.

Why should he write? Is it vanity that urges him? There is glory enough in the knife. Is it for money? One can make too much money. No. It is to search for some meaning in the ritual of surgery, which is at once murderous, painful, healing, and full of love. It is a devilish hard thing to transmit—to find, even. Perhaps if one were to cut out a heart, a lobe of the liver, a single convolution of the brain, and paste it to a page, it would speak with more eloquence than all the words of Balzac. Such a piece would need no literary style, no mass of erudition or history, but in its very shape and feel would tell all the frailty and strength, the despair and nobility of man. What? Publish a heart? A little piece of bone? Preposterous. Still I fear that is what it may require to reveal the truth that lies hidden in the body. Not all the undressings of Rabelais, Chekhov, or even William Carlos Williams have wrested it free, although God knows each one of those doctors made a heroic assault upon it.

I have come to believe that it is the flesh alone that counts. The rest is that with which we distract ourselves when we are not hungry or cold, in pain or ecstasy. In the recesses of the body I search for the philosophers' stone. I know it is there, hidden in the deepest, dampest cul-de-sac. It awaits discovery. To find it would be like the harnessing of fire. It would illuminate the world. Such a quest is not without pain. Who can gaze on so much misery and feel no hurt? Emerson has written that the poet is the only true doctor. I believe him, for the poet, lacking the impediment of speech with which the rest of us are afflicted, gazes, records, diagnoses, and prophesies.

I invited a young diabetic woman to the operating room to amputate her leg. She could not see the great shaggy black ulcer upon her foot and ankle that threatened to encroach upon the rest of her body, for she was blind as well. There upon her foot was a Mississippi Delta brimming with corruption, sending its raw tributaries down between her toes. Gone were all the little web spaces that when fresh and whole are such a delight to loving men. She could not see her wound, but she could feel it. There is no pain like that of the bloodless limb turned rotten and festering. There is neither unguent nor anodyne to kill such a pain yet leave intact the body.

For over a year I trimmed away the putrid flesh, cleansed, anointed, and dressed the foot, staving off, delaying. Three times each week, in her darkness, she sat upon my table, rocking back and forth, holding her extended leg by the thigh, gripping it as though it were a rocket that must be steadied lest it explode and scatter her toes about the room. And I would cut away a bit here, a bit there, of the swollen blue leather that was her tissue.

At last we gave up, she and I. We could no longer run ahead of the gangrene. We had not the legs for it. There must be an amputation in order that she might live—and I as well. It was to heal us both that I must take up knife and saw, and cut the leg off. And when I could feel it drop from her body to the table, see the blessed *space* appear between her and that leg, I too would be well.

Now it is the day of the operation. I stand by while the anesthetist administers the drugs, watch as the tense familiar body relaxes into narcosis. I turn then to uncover the leg. There, upon her kneecap, she has drawn, blindly, upside down for me to see, a face; just a circle with two ears, two eyes, a nose, and a smiling upturned mouth. Under it she has printed SMILE, DOCTOR. Minutes later I listen to the sound of the saw, until a little crack at the end tells me it is done.

So, I have learned that man is not ugly, but that he is Beauty itself. There is no other his equal. Are we not all dying, none faster or more slowly than any other? I have become receptive to the possibilities of love (for it is love, this thing that happens in the operating room), and each day I wait, trembling in the busy air. Perhaps today it will come. Perhaps today I will find it, take part in it, this love that blooms in the stoniest desert.

All through literature the doctor is portrayed as a figure of fun. Shaw was splenetic about him; Molière delighted in pricking his pompous medicine men, and well they deserved it. The doctor is ripe for caricature. But I believe that the truly great writing about doctors has not yet been done. I think it must be done *by* a doctor, one who is through with the love affair with his technique, who recognizes that he has played Narcissus, raining kisses on a mirror, and who now, out of the impacted masses of his guilt, has expanded into self-doubt, and finally into the high state of wonderment. Perhaps he will be a non-believer who, after a lifetime of grand gestures and mighty deeds, comes upon the knowledge that he has done no more than meddle in the lives of his fellows, and that he has done at least as much harm as good. Yet he may continue to pretend, at least, that there is nothing to fear, that death will not come, so long as people depend on his authority. Later, after his patients have left, he may closet himself in his darkened office, sweating and afraid.

There is a story by Unamuno in which a priest, living in a small Spanish village, is adored by all the people for his piety, kindness, and the majesty with which he celebrates the mass each Sunday. To them he is already a saint. It is a foregone conclusion, and they speak of him as Saint Immanuel. He helps them with their plowing and planting, tends them when they are sick, confesses them, comforts them in death, and every Sunday, in his rich, thrilling voice, transports them to paradise with his chanting. The fact is that Don Immanuel is not so much a saint as a martyr. Long ago his own faith left him. He is an atheist, a good man doomed to suffer the life of a hypocrite, pretending to a faith he does not have. As he raises the chalice of wine, his hands tremble, and a cold sweat pours from him. He cannot stop for he knows that the people need this of him, that their need is greater than his sacrifice. Still . . . still . . . could it be that Don Immanuel's whole life is a kind of prayer, a paean to God?

A writing doctor would treat men and women with equal reverence, for what is the "liberation" of either sex to him who knows the diagrams, the inner geographies of each? I love the solid heft of men as much as I adore the heated capaciousness of women—women in whose penetralia is found the repository of existence. I would have them glory in that. Women are physics and chemistry. They are matter. It is their bodies that tell of the frailty of men. Men have not their cellular, enzymatic wisdom. Man is albuminoid, proteinaceous, laked pearl; woman is yolky, ovoid, rich. Both are exuberant bloody growths. I would use the defects and deformities of each for my sacred purpose of writing, for I know that it is the marred and scarred and faulty that are subject to grace. I would seek the soul in the facts of animal economy and profligacy. Yes, it is the exact location of the soul that I am after. The smell of it is in my nostrils. I have caught glimpses of it in the body diseased. If only I could tell it. Is there no mathematical equation that

can guide me? So much pain and pus equals so much truth? It is elusive as the whippoorwill that one hears calling incessantly from out the night window, but which, nesting as it does low in the brush, no one sees. No one but the poet, for he sees what no one else can. He was born with the eye for it.

Once I thought I had it: Ten o'clock one night, the end room off a long corridor in a college infirmary, my last patient of the day, degree of exhaustion suitable for the appearance of a vision, some manifestation. The patient is a young man recently returned from Guatemala, from the excavation of Mayan ruins. His left upper arm wears a gauze dressing which, when removed, reveals a clean punched-out hole the size of a dime. The tissues about the opening are swollen and tense. A thin brownish fluid lips the edge, and now and then a lazy drop of the overflow spills down the arm. An abscess, inadequately drained. I will enlarge the opening to allow better egress of the pus. Nurse, will you get me a scalpel and some . . .?

What happens next is enough to lay Francis Drake avomit in his cabin. No explorer ever stared in wilder surmise than I into that crater from which there now emerges a narrow gray head whose sole distinguishing feature is a pair of black pincers. The head sits atop a longish flexible neck arching now this way, now that, testing the air. Alternately it folds back upon itself, then advances in new boldness. And all the while, with dreadful rhythmicity, the unspeakable pincers open and close. Abscess? Pus? Never. Here is the lair of a beast at whose malignant purpose I could but guess. A Mayan devil, I think, that would soon burst free to fly about the room, with horrid blanket-wings and iridescent scales, raking, pinching, injecting God knows what acid juice. And even now the irony does not escape me, the irony of my patient as excavator excavated.

With all the ritual deliberation of a high priest I advance a surgical clamp toward the hole. The surgeon's heart is become a bat hanging upside down from his rib cage. The rim achieved—now thrust—and the ratchets of the clamp close upon the empty air. The devil has retracted. Evil mocking laughter bangs back and forth in the brain. More stealth. Lying in wait. One must skulk. Minutes pass, perhaps an hour. . . . A faint disturbance in the lake, and once again the thing upraises, farther and farther, hovering. Acrouch, strung, the surgeon is one with his instrument; there is no longer any boundary between its metal and his flesh. They are joined in a single perfect tool of extirpation. It is just for this that he was born. Now—thrust—and clamp—and *yes*. Got him!

Transmitted to the fingers comes the wild thrashing of the creature. Pinned and wriggling, he is mine. I hear the dry brittle scream of the dragon, and a hatred seizes me, but such a detestation as would make of Iago a drooling sucktit. It is the demented hatred of the victor for the vanquished, the warden for his prisoner. It is the hatred of fear. Within the jaws of my hemostat is the whole of the evil of the world, the dark concentrate itself, and I shall kill it. For mankind. And, in so doing, will open the way into a thousand years of perfect peace. Here is Surgeon as Savior indeed.

Tight grip now . . . steady, relentless pull. How it scrabbles to keep its tentacle-hold. With an abrupt moist plop the extraction is complete. There, writhing in the teeth of the clamp, is a dirty gray body, the size and shape of an English walnut. He is hung everywhere with tiny black hooklets. Quickly . . . into the specimen jar of saline . . . the lid screwed tight. Crazily he swims round and round, wiping his slimy head against the glass, then slowly sinks to the bottom, the mass of hooks in frantic agonal wave.

"You are going to be all right," I say to my patient. "We are *all* going to be all right from now on."

The next day I take the jar to the medical school. "That's the larva of the botfly," says a pathologist. "The fly usually bites a cow and deposits its eggs beneath the skin. There, the egg develops into the larval form which, when ready, burrows its way to the outside through the hide and falls to the ground. In time it matures into a full-grown botfly. This one happened to bite a man. It was about to come out on its own, and, of course, it would have died."

The words *imposter, sorehead, servant of Satan* spring to my lips. But now he has been joined by other scientists. They nod in agreement. I gaze from one gray eminence to another, and know the mallet-blow of glory pulverized. I tried to save the world, but it didn't work out.

No, it is not the surgeon who is God's darling. He is the victim of vanity. It is the poet who heals with his words, stanches the flow of blood, stills the rattling breath, applies poultice to the scalded flesh.

Did you ask me why a surgeon writes? I think it is because I wish to be a doctor.

Name ______________________ **Course** ____________ **Date** ______________

Chapter Questions

1. Why do you think the author is so disappointed when a colleague easily dismisses the threat of the "creature" he removed in a young patient?

2. Doctors use education, instruments, and medicine to save lives, but what do you think the author would say is the true purpose of a doctor?

23-YEAR OLD KIM SUOZZI UNDERGOES CRYONIC PRESERVATION AFTER SUCCESSFUL FUNDRAISING CAMPAIGN

George Dvorsky

The science of Cryonics has been fodder for science fiction and fantasy writers for years. The main philosophy of this technology is preserving and storing the human body in the event of a possible future cure for the ailment that took the life thus restoring the person back to livelihood at a later date. This remote possibility has convinced some hopeful individuals to choose cryonic suspension. George Dvorsky is a Canadian born bioethicist, transhumanist, and futurist. He currently serves as chair of the board of the Institute for Ethics and Emerging Technologies (IEET). He is a contributor to IO9, a blog focusing on the subjects of science, technology, futurism, science fiction and fantasy. This story highlights efforts made by 23-year-old neuroscience student, Kim Suozzi, who was diagnosed with terminal brain cancer, and her fundraising efforts for her own cryonic preservation. As of March 31, 2016, Alcor Life Extension Foundation in Scottsdale, Arizona, had 146 "patients" in cryonic suspension awaiting the possibility of being restored to life at a later date. Over 1000 other individuals have become Alcor members, therefore pledging their bodies to be frozen and preserved after death in the hope that, in the future, they may be brought back to life.

© photosoft/Shutterstock.com

Back in September we told you about Kim Suozzi, the 23-year old neuroscience student who was in the midst of battling terminal brain cancer. Sadly, she passed away last week on January 17th—but not before a successful fundraising campaign managed to secure the funds required to grant Suozzi her dying wish: cryonic preservation.

Suozzi's effort to get cryonically preserved began when she asked the Reddit community for help. At the time, she was only expected to live for another three to six months, making her request for financial assistance all the more urgent.

"I want to be cryogenically preserved when I die from brain cancer but can't afford it," she wrote, "I am literally begging for financial help."

Needless to say, preservation and storage at a cryonics facility like Alcor Life Extension Foundation or Cryonics Institute is not cheap. A standard Alcor suspension costs $70,000.00, which includes high quality stand-by, neuropreservation, and storage—but only if the terminal member relocates to the Phoenix area (otherwise the additional stand-by and transport costs bring the total to $80,000.00).

A whole-body preservation at Alcor costs $200,000.00. Cryonics Institute charges $28,000.00 for a whole-body preservation, but this fee does not include stand-by and transportation costs.

Both companies use vitrification preservation chemicals that stops the formation of damaging ice crystals, but they also have different facilities and procedures.

Most members of Alcor and CI are able to meet these costs by taking out a second life insurance policy and naming the cryonics company of their choice as the beneficiary. Preservations are only conducted on patients who have already been pronounced clinically dead.

© anyaivanova/Shutterstock.com

Soon after Suozzi's posting on Reddit, she set up her own fundraising campaign.

Quickly thereafter, a cryonics-friendly futurist group called Society for Venturism set up its own charitable fund. This group, a volunteer-run not-for-profit, is no stranger to this process; it has already successfully raised funds for two cryopreservations and is currently working on one urgent case.

Between the two fundraising efforts, enough money was raised for Suozzi to be cryonically preserved at Alcor—a procedure that was undertaken a few days ago. Alcor will be issuing a statement shortly.

"When funds are raised for a cryonics charity recipient they can choose to contract with the cryonics organization of their choice," Shannon Vyff told io9. "It is up to them to set up their contract; the Society for Venturism will then provide funds to the cryonics organization they have contracted with." Vyff is a board of Society for Venturism, and an Alcor and Cryonics Institute member.

"I have been happy to help the cryonics community, and at times it is hard raise funds for a charity recipient," she told us. "Kim's case was compelling to many people—not only did many cryonicists donate but non-cryonicists as well." After having worked on one charity case before and hearing about how two past cases went, Vyff was impressed with how quickly funds were raised for Suozzi. Within just months she had enough to fund a neurosuspension with Alcor.

Speaking through an official announcement, Suozzi's boyfriend had this to say:

Our hope is that technology will continue to progress to the point that Kim may have a real chance of living again in the future. Unfortunately, the development of the requisite technologies could be decades or centuries away. Since Kim is no longer with us to explore and innovate in the field of neuroscience, she is counting on all of us to push for the innovations she had hoped to see in her lifetime.

Until (or unless) the day comes that Kim can be brought back, remember her, celebrate her, and emulate her resilience, so we can create the future of her dreams.

Nobody is too young to make cryopreservation arrangements.

Prior to her death, Suozzi had been in touch with another Society for Venturism case, Aaron Winborn, a 45-year-old man struggling with ALS. Hoping to help, but frustrated by her failing health, she contacted him and apologized that she could not do more.

Name ______________________ Course ____________ Date ______________

Chapter Questions

1. Define Cryonics or Cryonic suspension.

2. According to this article, what is the largest challenge to freezing human tissue? Is there a possible solution to this problem?

3. What are the religious and moral objections to this science in this story?

4. How do you feel about this process?

SECTION 6 The Perils of Technology and What the Future Holds

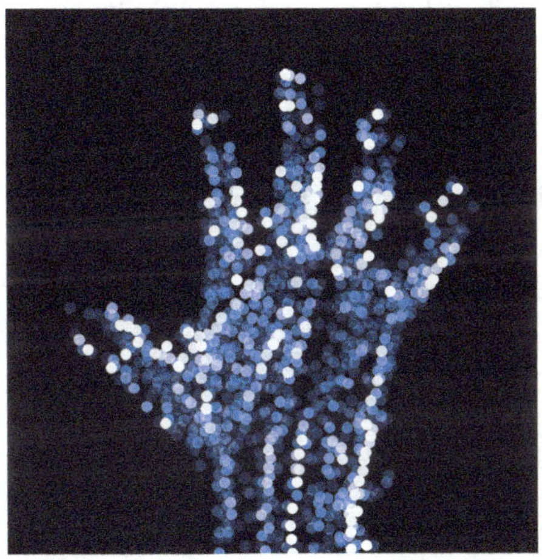

© majcot/Shutterstock.com

Introduction

It can be argued that some technologies such as antibiotics, genetically modified foods (GMOs), and social media have been beneficial to society. After all, how many people haven't been cured of a bacterial infection with modern medicine or reconnected with an old friend through social media? Agricultural advancements are feeding more people now with genetically modified foods, and the newest, fastest computers allow us to communicate more efficiently in the workplace and locate that hard to find item in seconds.

But, are there drawbacks? For every movement forward, is there a cost? A sick person may be cured today with antibiotics; however is the human body becoming more and more resistant to them, creating a need for even stronger medicines? Have these given rise to Superbugs? Similarly, it can be argued that Facebook also has negatively impacted society. For instance, according to the *New York Post*, one-third of divorce proceedings today mention the word "Facebook" within its citings. Numerous articles are also available with job loss statistics based on Facebook user postings, particularly photographs.

Not all benefits and drawbacks are modern day. In August of 1492, Christopher Columbus set sail with three ships and eighty-eight men. His ships were caravels, the new and more maneuverable sea vessel developed in the fifteenth century by the sea-faring Portuguese. On his voyage, he landed on and explored what is known today as Samana Cay. His contemporaries soon explored and claimed the region today known as Florida as well as modern-day Mexico, leading to a blend of Native American and European cultures, much of which is still evident today. However, did this exploration (due largely to the efficiency of the caravels), its findings, and subsequent history turn out successful for everyone involved? No. The Native American peoples were not resistant to European diseases. Some reports show that as many as ninety percent of the native population was wiped out due to European disease introduction.

This list can and does go on. As a student reading through this text ask yourself these questions about each technology studied: What are the benefits of this technology? Do these benefits come with a price? Will the perils be seen in the future? Do the benefits outweigh these drawbacks? Finally, consider whether the technology itself is bad or if the problems come with humans' use, or misuse, of it.

When researching the benefits and drawbacks of technologies, it is also important to look forward toward the future. For example, in 1964, Gene Roddenberry created the concept for the show *Star Trek*. His characters explored space with the assistance of advanced gadgets such as phasers, communicators, telepresence, and the universal translator. By 2015, these and other items showcased in the TV series has come to fruition in some form. In a mere 50 years these items went from science fiction to reality.

One can just as easily wonder what the next fifty years will bring. No doubt the digital age will shape future advancements. As stated above, humans will be tasked with deciding how to use new advancements, and whether technology will be used to advance or ultimately harm society.

THE UNKNOWN CITIZEN

W. H. Auden

W. H. Auden's "The Unknown Citizen" reveals a future in which people have been reduced to numbers and statistics. The technological ability to store and gather the information has led to that information overshadowing individual identity. The excerpt from The Axemaker's Gift *in the Communication chapter introduces the Cut and Control Theory which holds that a society's elites are able to exert a measure of control over the common citizen via the numbers used to track and sort them in a complex society. How might this poem illustrate that?*

© ktsdesign/Shutterstock.com

[To JS/07/M/378 This Marble Monument Is Erected by the State]
He was found by the Bureau of Statistics to be
One against whom there was no official complaint,
And all the reports on his conduct agree
That, in the modern sense of an old-fashioned word, he was a saint,
For in everything he did he served the Greater Community.
Except for the War till the day he retired
He worked in a factory and never got fired,
But satisfied his employers, Fudge Motors Inc.
Yet he wasn't a scab or odd in his views,

For his Union reports that he paid his dues,
(Our report on his Union shows it was sound)
And our Social Psychology workers found
That he was popular with his mates and liked a drink.
The Press are convinced that he bought a paper every day
And that his reactions to advertisements were normal in every way.
Policies taken out in his name prove that he was fully insured,
And his Health-card shows he was once in hospital but left it cured.
Both Producers Research and High-Grade Living declare
He was fully sensible to the advantages of the Installment Plan
And had everything necessary to the Modern Man,
A phonograph, a radio, a car and a frigidaire.
Our researchers into Public Opinion are content
That he held the proper opinions for the time of year;
When there was peace, he was for peace, when there was war, he went.
He was married and added five children to the population,
Which our Eugenist says was the right number of a parent of his generation,
And our teachers report that he never interfered with their education.
Was he free? Was he happy? The question is absurd:
Had anything been wrong, we should certainly have heard.

Name ______________________ **Course** ______________ **Date** ________________

Chapter Questions

1. Describe how this poem is satirical.

2. How do the last two lines of this poem reveal its central theme?

THE VELDT

Ray Bradbury

Ray Bradbury, author of "The Veldt" and "There Will Come Soft Rains," continues to inspire generations of readers to think, dream and create. He wrote hundreds of short stories, fifty books, as well as numerous essays, poems and screen plays with genres including fantasy, science fiction, and horror. He received several distinguished awards including the 2000 National Book Foundation Medal for Distinguished Contribution to American Letters, the 2004 National Medal of Arts, and the 2007 Pulitzer Prize Special Citation. A completely automated house and a nursery, where children explore any location they choose, are major components of his writings in this unit. Both written in 1950, the stories explore the idea of virtual reality making them at least five decades ahead of their time.

© Johan Swanepoel/Shutterstock.com

"George, I wish you'd look at the nursery."

"What's wrong with it?"

"I don't know."

"Well, then."

"I just want you to look at it, is all, or call a psychologist in to look at it."

"What would a psychologist want with a nursery?"

"You know very well what he'd want." His wife paused in the middle of the kitchen and watched the stove busy humming to itself, making supper for four.

"It's just that the nursery is different now than it was."

"All right, let's have a look."

They walked down the hall of their soundproofed Happylife Home, which had cost them thirty thousand dollars installed, this house which clothed and fed and rocked them to sleep and played and sang and was good to them.

Their approach sensitized a switch somewhere and the nursery light flicked on when they came within ten feet of it. Similarly, behind them, in the halls, lights went on and off as they left them behind, with a soft automaticity.

"Well," said George Hadley.

They stood on the thatched floor of the nursery. It was forty feet across by forty feet long and thirty feet high; it had cost half again as much as the rest of the house. "But nothing's too good for our children," George had said.

The nursery was silent. It was empty as a jungle glade at hot high noon. The walls were blank and two dimensional. Now, as George and Lydia Hadley stood in the center of the room, the walls began to purr and recede into crystalline distance, it seemed, and presently an African veldtappeared, in three dimensions, on all sides, in color reproduced to the final pebble and bit of straw. The ceiling above them became a deep sky with a hot yellow sun.

George Hadley felt the perspiration start on his brow.

"Let's get out of this sun," he said. "This is a little too real. But I don't see anything wrong."

"Wait a moment, you'll see," said his wife.

Now the hidden odorophonics were beginning to blow a wind of odor at the two people in the middle of the baked veldtland. The hot straw smell of lion grass, the cool green smell of the hidden water hole, the great rusty smell of animals, the smell of dust like a red paprika in the hot air. And now the sounds: the thump of distant antelope feet on grassy sod, the paperyrustling of vultures. A shadow passed through the sky. The shadow flickered on George Hadley's upturned, sweating face.

"Filthy creatures," he heard his wife say.

"The vultures."

"You see, there are the lions, far over, that way. Now they're on their way to the water hole. They've just been eating," said Lydia. "I don't know what."

"Some animal." George Hadley put his hand up to shield off the burning light from his squinted eyes. "A zebra or a baby giraffe, maybe."

"Are you sure?" His wife sounded peculiarly tense.

"No, it's a little late to be sure," he said, amused. "Nothing over there I can see but cleaned bone, and the vultures dropping for what's left."

"Did you hear that scream?" she asked.

"No."

"About a minute ago?"

"Sorry, no."

The lions were coming. And again George Hadley was filled with admiration for the mechanical genius who had conceived this room. A miracle of efficiency selling for an absurdly low price. Every home should have one.

Oh, occasionally they frightened you with their clinical accuracy, they startled you, gave you a twinge, but most of the time what fun for everyone, not only your own son and daughter, but for yourself when you felt like a quick jaunt to a foreign land, a quick change of scenery. Well, here it was!

And here were the lions now, fifteen feet away, so real, so feverishly and startlingly real that you could feel the prickling fur on your hand, and your mouth was stuffed with the dusty upholstery smell of their heated pelts, and the yellow of them was in your eyes like the yellow of an exqui site French tapestry, the yellows of lions and summer grass, and the sound of the matted lion lungs exhaling on the silent noontide, and the smell of meat from the panting, dripping mouths.

The lions stood looking at George and Lydia Hadley with terrible green-yellow eyes.

"Watch out!" screamed Lydia.

The lions came running at them.

Lydia bolted and ran. Instinctively, George sprang after her. Outside, in the hall, with the door slammed he was laughing and she was crying, and they both stood appalled at the other's reaction.

"George!"

"Lydia! Oh, my dear poor sweet Lydia!"

"They almost got us!"

"Walls, Lydia, remember; crystal walls, that's all they are. Oh, they look real, I must admit—Africa in your parlor—but it's all dimensional,superreactionary, supersensitive color film and mental tape film behind glass screens. It's all odorophonics and sonics, Lydia. Here's my handkerchief."

"I'm afraid." She came to him and put her body against him and cried steadily. "Did you see? Did you feel? It's too real."

"Now, Lydia . . ."

"You've got to tell Wendy and Peter not to read any more on Africa."

"Of course—of course." He patted her.

"Promise?"

"Sure."

"And lock the nursery for a few days until I get my nerves settled."

"You know how difficult Peter is about that. When I punished him a month ago by locking the nursery for even a few hours—the tantrum be threw! And Wendy too. They live for the nursery."

"It's got to be locked, that's all there is to it."

"All right." Reluctantly he locked the huge door. "You've been working too hard. You need a rest."

"I don't know—I don't know," she said, blowing her nose, sitting down in a chair that immediately began to rock and comfort her. "Maybe I don't have enough to do. Maybe I have time to think too much. Why don't we shut the whole house off for a few days and take a vacation?"

"You mean you want to fry my eggs for me?"

"Yes." She nodded.

"And darn my socks?"

"Yes." A frantic, watery-eyed nodding.

"And sweep the house?"

"Yes, yes—oh, yes!"

"But I thought that's why we bought this house, so we wouldn't have to do anything?"

"That's just it. I feel like I don't belong here. The house is wife and mother now, and nursemaid. Can I compete with an African veldt? Can I give a bath and scrub the children as efficiently or quickly as the automatic scrub bath can? I cannot. And it isn't just me. It's you. You've been awfully nervous lately."

"I suppose I have been smoking too much."

"You look as if you didn't know what to do with yourself in this house, either. You smoke a little more every morning and drink a little more every afternoon and need a little more sedative every night. You're beginning to feel unnecessary too."

"Am I?" He paused and tried to feel into himself to see what was really there.

"Oh, George!" She looked beyond him, at the nursery door. "Those lions can't get out of there, can they?"

He looked at the door and saw it tremble as if something had jumped against it from the other side.

"Of course not," he said.

At dinner they ate alone, for Wendy and Peter were at a special plastic carnival across town and had televised home to say they'd be late, to go ahead eating. So George Hadley, bemused, sat watching the dining-room table produce warm dishes of food from its mechanical interior.

"We forgot the ketchup," he said.

"Sorry," said a small voice within the table, and ketchup appeared.

As for the nursery, thought George Hadley, it won't hurt for the children to be locked out of it awhile. Too much of anything isn't good for anyone. And it was clearly indicated that the children had been spending a little too much time on Africa. That sun. He could feel it on his neck, still, like a hot paw. And the lions. And the smell of blood. Remarkable how the nursery caught the telepathic emanations of the children's minds and created life to fill their every desire. The children thought lions, and there were lions. The children thought zebras, and there were zebras. Sun—sun. Giraffes—giraffes. Death and death.

That last. He chewed tastelessly on the meat that the table had cut for him. Death thoughts. They were awfully young, Wendy and Peter, for death thoughts. Or, no, you were never too young, really. Long before you knew what death was you were wishing it on someone else. When you were two years old you were shooting people with cap pistols.

But this—the long, hot African veldt-the awful death in the jaws of a lion. And repeated again and again.

"Where are you going?"

He didn't answer Lydia. Preoccupied, be let the lights glow softly on ahead of him, extinguish behind him as he padded to the nursery door. He listened against it. Far away, a lion roared.

He unlocked the door and opened it. Just before he stepped inside, he heard a faraway scream. And then another roar from the lions, which subsided quickly.

He stepped into Africa. How many times in the last year had he opened this door and found Wonderland, Alice, the Mock Turtle, or Aladdin and his Magical Lamp, or Jack Pumpkin head of Oz, or Dr. Doolittle, or the cow jumping over a very real-appearing moon-all the delightful contraptions of a make-believe world. How often had he seen Pegasus flying in the sky ceiling, or seen fountains of red fireworks, or heard angel voices singing. But now, is yellow hot Africa, this bake oven with murder in the heat. Perhaps Lydia was right. Perhaps they needed a little vacation from the fantasy which was growing a bit too real for ten-year-old children. It was all right to exercise one's mind with gymnastic fantasies, but when the lively child mind settled on one pattern . . .? It seemed that, at a distance, for the past month, he had heard lions roaring, and smelled their strong odor seeping as far away as his study door. But, being busy, he had paid it no attention.

George Hadley stood on the African grassland alone. The lions looked up from their feeding, watching him. The only flaw to the illusion was the open door through which he could see his wife, far down the dark hall, like a framed picture, eating her dinner abstractedly.

"Go away," he said to the lions.

They did not go.

He knew the principle of the room exactly. You sent out your thoughts.

Whatever you thought would appear. "Let's have Aladdin and his lamp," he snapped. The veldtland remained; the lions remained.

"Come on, room! I demand Aladdin!" he said.

Nothing happened. The lions mumbled in their baked pelts.

"Aladdin!"

He went back to dinner. "The fool room's out of order," he said. "It won't respond."

"Or—"

"Or what?"

"Or it can't respond," said Lydia, "because the children have thought about Africa and lions and killing so many days that the room's in a rut."

"Could be."

"Or Peter's set it to remain that way."

"Set it?"

"He may have got into the machinery and fixed something."

"Peter doesn't know machinery."

He's a wise one for ten. That I.Q. of his—"

"Nevertheless—"

"Hello, Mom. Hello, Dad."

The Hadleys turned. Wendy and Peter were coming in the front door, cheeks like peppermint candy, eyes like bright blue agate marbles, a smell of ozone on their jumpers from their trip in the helicopter.

"You're just in time for supper," said both parents.

"We're full of strawberry ice cream and hot dogs," said the children, holding hands. "But we'll sit and watch."

"Yes, come tell us about the nursery," said George Hadley.

The brother and sister blinked at him and then at each other.

"Nursery?"

"All about Africa and everything," said the father with false joviality.

"I don't understand," said Peter.

"Your mother and I were just traveling through Africa with rod and reel; Tom Swift and his Electric Lion," said George Hadley.

"There's no Africa in the nursery," said Peter simply.

"Oh, come now, Peter. We know better."

"I don't remember any Africa," said Peter to Wendy. "Do you?"

"No."

"Run see and come tell."

She obeyed

"Wendy, come back here!" said George Hadley, but she was gone. The house lights followed her like a flock of fireflies. Too late, he realized he had forgotten to lock the nursery door after his last inspection.

"Wendy'll look and come tell us," said Peter.

"She doesn't have to tell me. I've seen it."

"I'm sure you're mistaken, Father."

"I'm not, Peter. Come along now."

But Wendy was back. "It's not Africa," she said breathlessly.

"We'll see about this," said George Hadley, and they all walked down the hall together and opened the nursery door.

There was a green, lovely forest, a lovely river, a purple mountain, high voices singing, and Rima, lovely and mysterious, lurking in the trees with colorful flights of butterflies, like animated bouquets, lingering in her long hair. The African veldtland was gone. The lions were gone. Only Rima was here now, singing a song so beautiful that it brought tears to your eyes.

George Hadley looked in at the changed scene. "Go to bed," he said to the children.

They opened their mouths.

"You heard me," he said.

They went off to the air closet, where a wind sucked them like brown leaves up the flue to their slumber rooms.

George Hadley walked through the singing glade and picked up something that lay in the comer near where the lions had been. He walked slowly back to his wife.

"What is that?" she asked.

"An old wallet of mine," he said.

He showed it to her. The smell of hot grass was on it and the smell of a lion. There were drops of saliva on it, it had been chewed, and there were blood smears on both sides.

He closed the nursery door and locked it, tight.

In the middle of the night he was still awake and he knew his wife was awake. "Do you think Wendy changed it?" she said at last, in the dark room.

"Of course."

"Made it from a veldt into a forest and put Rima there instead of lions?"

"Yes."

"Why?"

"I don't know. But it's staying locked until I find out."

"How did your wallet get there?"

"I don't know anything," he said, "except that I'm beginning to be sorry we bought that room for the children. If children are neurotic at all, a room like that—"

"It's supposed to help them work off their neuroses in a healthful way."

"I'm starting to wonder." He stared at the ceiling.

"We've given the children everything they ever wanted. Is this our reward-secrecy, disobedience?"

"Who was it said, 'Children are carpets, they should be stepped on occasionally'? We've never lifted a hand. They're insufferable—let's admit it. They come and go when they like; they treat us as if we were offspring. They're spoiled and we're spoiled."

"They've been acting funny ever since you forbade them to take the rocket to New York a few months ago."

"They're not old enough to do that alone, I explained."

"Nevertheless, I've noticed they've been decidedly cool toward us since."

"I think I'll have David McClean come tomorrow morning to have a look at Africa."

"But it's not Africa now, it's Green Mansions country and Rima."

"I have a feeling it'll be Africa again before then."

A moment later they heard the screams.

Two screams. Two people screaming from downstairs. And then a roar of lions.

"Wendy and Peter aren't in their rooms," said his wife.

He lay in his bed with his beating heart. "No," he said. "They've broken into the nursery."

"Those screams—they sound familiar."

"Do they?"

"Yes, awfully."

And although their beds tried very hard, the two adults couldn't be rocked to sleep for another hour. A smell of cats was in the night air.

"Father?" said Peter.

"Yes."

Peter looked at his shoes. He never looked at his father any more, nor at his mother. "You aren't going to lock up the nursery for good, are you?"

"That all depends."

"On what?" snapped Peter.

"On you and your sister. If you intersperse this Africa with a little variety—oh, Sweden perhaps, or Denmark or China—"

"I thought we were free to play as we wished."

"You are, within reasonable bounds."

"What's wrong with Africa, Father?"

"Oh, so now you admit you have been conjuring up Africa, do you?"

"I wouldn't want the nursery locked up," said Peter coldly. "Ever."

"Matter of fact, we're thinking of turning the whole house off for about a month. Live sort of a carefree one-for-all existence."

"That sounds dreadful! Would I have to tie my own shoes instead of letting the shoe tier do it? And brush my own teeth and comb my hair and give myself a bath?"

"It would be fun for a change, don't you think?"

"No, it would be horrid. I didn't like it when you took out the picture painter last month."

"That's because I wanted you to learn to paint all by yourself, son."

"I don't want to do anything but look and listen and smell; what else is there to do?"

"All right, go play in Africa."

"Will you shut off the house sometime soon?"

"We're considering it."

"I don't think you'd better consider it any more, Father."

"I won't have any threats from my son!"

"Very well." And Peter strolled off to the nursery.

"Am I on time?" said David McClean.

"Breakfast?" asked George Hadley.

"Thanks, had some. What's the trouble?"

"David, you're a psychologist."

"I should hope so."

"Well, then, have a look at our nursery. You saw it a year ago when you dropped by; did you notice anything peculiar about it then?"

"Can't say I did; the usual violences, a tendency toward a slight paranoia here or there, usual in children because they feel persecuted by parents constantly, but, oh, really nothing."

They walked down the hall. "I locked the nursery up," explained the father, "and the children broke back into it during the night. I let them stay so they could form the patterns for you to see."

There was a terrible screaming from the nursery.

"There it is," said George Hadley. "See what you make of it."

They walked in on the children without rapping.

The screams had faded. The lions were feeding.

"Run outside a moment, children," said George Hadley. "No, don't change the mental combination. Leave the walls as they are. Get!"

With the children gone, the two men stood studying the lions clustered at a distance, eating with great relish whatever it was they had caught.

"I wish I knew what it was," said George Hadley. "Sometimes I can almost see. Do you think if I brought high-powered binoculars here and—"

David McClean laughed dryly. "Hardly." He turned to study all four walls. "How long has this been going on?"

"A little over a month."

"It certainly doesn't feel good."

"I want facts, not feelings."

"My dear George, a psychologist never saw a fact in his life. He only hears about feelings; vague things. This doesn't feel good, I tell you.

Trust my hunches and my instincts. I have a nose for something bad. This is very bad. My advice to you is to have the whole damn room torn down and your children brought to me every day during the next year for treatment."

"Is it that bad?"

"I'm afraid so. One of the original uses of these nurseries was so that we could study the patterns left on the walls by the child's mind, study at our leisure, and help the child. In this case, however, the room has become a channel toward-destructive thoughts, instead of a release away from them."

"Didn't you sense this before?"

"I sensed only that you bad spoiled your children more than most. And now you're letting them down in some way. What way?"

"I wouldn't let them go to New York."

"What else?"

"I've taken a few machines from the house and threatened them, a month ago, with closing up the nursery unless they did their homework. I did close it for a few days to show I meant business."

"Ah, ha!"

"Does that mean anything?"

"Everything. Where before they had a Santa Claus now they have a Scrooge. Children prefer Santas. You've let this room and this house replaceyou and your wife in your children's affections. This room is their mother and father, far more important in their lives than their real parents. And now you come along and want to shut it off. No wonder there's hatred here.

You can feel it coming out of the sky. Feel that sun. George, you'll have to change your life. Like too many others, you've built it around creature comforts. Why, you'd starve tomorrow if something went wrong in your kitchen. You wouldn't know bow to tap an egg. Nevertheless, turn everything off. Start new. It'll take time. But we'll make good children out of bad in a year, wait and see."

"But won't the shock be too much for the children, shutting the room upabruptly, for good?"

"I don't want them going any deeper into this, that's all."

The lions were finished with their red feast.

The lions were standing on the edge of the clearing watching the two men.

"Now I'm feeling persecuted," said McClean. "Let's get out of here. I never have cared for these damned rooms. Make me nervous."

"The lions look real, don't they?" said George Hadley. I don't suppose there's any way—"

"What?"

"—that they could become real?"

"Not that I know."

"Some flaw in the machinery, a tampering or something?"

"No."

They went to the door.

"I don't imagine the room will like being turned off," said the father.

"Nothing ever likes to die—even a room."

"I wonder if it hates me for wanting to switch it off?"

"Paranoia is thick around here today," said David McClean. "You can follow it like a spoor. Hello." He bent and picked up a bloody scarf. "This yours?"

"No." George Hadley's face was rigid. "It belongs to Lydia."

They went to the fuse box together and threw the switch that killed the nursery.

The two children were in hysterics. They screamed and pranced and threw things. They yelled and sobbed and swore and jumped at the furniture.

"You can't do that to the nursery, you can't!"

"Now, children."

The children flung themselves onto a couch, weeping.

"George," said Lydia Hadley, "turn on the nursery, just for a few moments. You can't be so abrupt."

"No."

"You can't be so cruel . . ."

"Lydia, it's off, and it stays off. And the whole damn house dies as of here and now. The more I see of the mess we've put ourselves in, the more it sickens me. We've been contemplating our mechanical, electronic navels for too long. My God, how we need a breath of honest air!"

And he marched about the house turning off the voice clocks, the stoves, the heaters, the shoe shiners, the shoe lacers, the body scrubbers and swabbers and massagers, and every other machine be could put his hand to.

The house was full of dead bodies, it seemed. It felt like a mechanical cemetery. So silent. None of the humming hidden energy of machines waiting to function at the tap of a button.

"Don't let them do it!" wailed Peter at the ceiling, as if he was talking to the house, the nursery. "Don't let Father kill everything." He turned to his father. "Oh, I hate you!"

"Insults won't get you anywhere."

"I wish you were dead!"

"We were, for a long while. Now we're going to really start living. Instead of being handled and massaged, we're going to live."

Wendy was still crying and Peter joined her again. "Just a moment, just one moment, just another moment of nursery," they wailed.

"Oh, George," said the wife, "it can't hurt."

"All right—all right, if they'll just shut up. One minute, mind you, and then off forever."

"Daddy, Daddy, Daddy!" sang the children, smiling with wet faces.

"And then we're going on a vacation. David McClean is coming back in half an hour to help us move out and get to the airport. I'm going to dress.

You turn the nursery on for a minute, Lydia, just a minute, mind you."

And the three of them went babbling off while he let himself be vacuumed upstairs through the air flue and set about dressing himself. A minute later Lydia appeared.

"I'll be glad when we get away," she sighed.

"Did you leave them in the nursery?"

"I wanted to dress too. Oh, that horrid Africa. What can they see in it?"

"Well, in five minutes we'll be on our way to Iowa. Lord, how did we ever get in this house? What prompted us to buy a nightmare?"

"Pride, money, foolishness."

"I think we'd better get downstairs before those kids get engrossed with those damned beasts again."

Just then they heard the children calling, "Daddy, Mommy, come quick—quick!"

They went downstairs in the air flue and ran down the hall. The children were nowhere in sight. "Wendy? Peter!"

They ran into the nursery. The veldtland was empty save for the lions waiting, looking at them. "Peter, Wendy?"

The door slammed.

"Wendy, Peter!"

George Hadley and his wife whirled and ran back to the door.

"Open the door!" cried George Hadley, trying the knob. "Why, they've locked it from the outside! Peter!" He beat at the door. "Open up!"

He heard Peter's voice outside, against the door.

"Don't let them switch off the nursery and the house," he was saying.

Mr. and Mrs. George Hadley beat at the door. "Now, don't be ridiculous, children. It's time to go. Mr. McClean'll be here in a minute and . . ."

And then they heard the sounds.

The lions on three sides of them, in the yellow veldt grass, padding through the dry straw, rumbling and roaring in their throats.

The lions.

Mr. Hadley looked at his wife and they turned and looked back at the beasts edging slowly forward crouching, tails stiff.

Mr. and Mrs. Hadley screamed.

And suddenly they realized why those other screams bad sounded familiar.

"Well, here I am," said David McClean in the nursery doorway, "Oh, hello." He stared at the two children seated in the center of the open glade eating a little picnic lunch. Beyond them was the water hole and the yellow veldtland; above was the hot sun. He began to perspire. "Where are your father and mother?"

The children looked up and smiled. "Oh, they'll be here directly."

"Good, we must get going." At a distance Mr. McClean saw the lions fighting and clawing and then quieting down to feed in silence under the shady trees.

He squinted at the lions with his hand tip to his eyes.

Now the lions were done feeding. They moved to the water hole to drink.

A shadow flickered over Mr. McClean's hot face. Many shadows flickered.

The vultures were dropping down the blazing sky.

"A cup of tea?" asked Wendy in the silence.

Name ______________________ **Course** ______________ **Date** ________________

Chapter Questions

1. Who is to blame for what happens at the end of the story, the parents or the children? Explain your answer.

2. How do you think the author views technology?

THERE WILL COME SOFT RAINS

Ray Bradbury

© winui/Shutterstock.com

In the living room the voice-clock sang, *Tick-tock, seven o'clock, time to get up, time to get up, seven o'clock!* as if it were afraid that nobody would. The morning house lay empty. The clock ticked on, repeating and repeating its sounds into the emptiness. *Seven-nine, breakfast time, seven-nine!*

In the kitchen the breakfast stove gave a hissing sigh and ejected from its warm interior eight pieces of perfectly browned toast, eight eggs sunnyside up, sixteen slices of bacon, two coffees, and two cool glasses of milk.

"Today is August 4, 2026," said a second voice from the kitchen ceiling, "in the city of Allendale, California." It repeated the date three times for memory's sake. "Today is Mr. Featherstone's birthday. Today is the anniversary of Tilita's marriage. Insurance is payable, as are the water, gas, and light bills."

Somewhere in the walls, relays clicked, memory tapes glided under electric eyes.

Eight-one, tick-tock, eight-one o'clock, off to school, off to work, run, run, eight-one! But no doors slammed, no carpets took the soft tread of rubber heels. It was raining outside. The weather box on the front door sang quietly: "Rain, rain, go away; rubbers, raincoats for today . . ." And the rain tapped on the empty house, echoing.

Outside, the garage chimed and lifted its door to reveal the waiting car. After a long wait the door swung down again.

At eight-thirty the eggs were shriveled and the toast was like stone. An aluminum wedge scraped them into the sink, where hot water whirled them down a metal throat which digested and flushed them away to the distant sea. The dirty dishes were dropped into a hot washer and emerged twinkling dry.

Nine-fifteen, sang the clock, *time to clean.*

Out of warrens in the wall, tiny robot mice darted. The room were acrawl with the small cleaning animals, all rubber and metal. They thudded against chairs, whirling their mustached runners, kneading the rug nap, sucking gently at hidden dust. Then, like mysterious invaders, they popped into their burrows. Their pink electric eyes faded. The house was clean.

Ten o'clock. The sun came out from behind the rain. The house stood alone in a city of rubble and ashes. This was the one house left standing. At night the ruined city gave off a radioactive glow which could be seen for miles.

Ten-fifteen. The garden sprinklers whirled up in golden founts, filling the soft morning air with scatterings of brightness. The water pelted windowpanes, running down the charred west side where the house had been burned evenly free of its white paint. The entire west face of the house was black, save for five places. Here the silhouette in paint of a man mowing a lawn. Here, as in a photograph, a woman bent to pick flowers. Still farther over, their images burned on wood in one titanic instant, a small boy, hands flung into the air; higher up, the image of a thrown ball, and opposite him a girl, hands raised to catch a ball which never came down.

The five spots of paint—the man, the woman, the children, the ball—remained. The rest was a thin charcoaled layer.

The gentle sprinkler rain filled the garden with falling light.

Bobboz/Shutterstock.com

Until this day, how well the house had kept its peace. How carefully it had inquired, "Who goes there? What's the password?" and, getting no answer from lonely foxes and whining cats, it had shut up its windows and drawn shades in an old-maidenly preoccupation with self-protection which bordered on a mechanical paranoia.

It quivered at each sound, the house did. If a sparrow brushed a window, the shade snapped up. The bird, startled, flew off! No, not even a bird must touch the house!

The house was an altar with ten thousand attendants, big, small, servicing, attending, in choirs. But the gods had gone away, and the ritual of the religion continued senselessly, uselessly.

Twelve noon.

A dog whined, shivering, on the front porch.

The front door recognized the dog voice and opened. The dog, once huge and fleshy, but now gone to bone and covered with sores, moved in and through the house, tracking mud. Behind it whirred angry mice, angry at having to pick up mud, angry at inconvenience.

For not a leaf fragment blew under the door but what the wall panels flipped open and the copper scrap rats flashed swiftly out. The offending dust, hair, or paper, seized in miniature steel jaws, was raced back to the burrows. There, down tubes which fed into the cellar, it was dropped into the sighing vent of an incinerator which sat like evil Baal in a dark corner.

The dog ran upstairs, hysterically yelping to each door, at last realizing, as the house realized, that only silence was here.

It sniffed the air and scratched the kitchen door. Behind the door, the stove was making pancakes which filled the house with a rich baked odor and the scent of maple syrup.

The dog frothed at the mouth, lying at the door, sniffing, its eyes turned to fire. It ran wildly in circles, biting at its tail, spun in a frenzy and died. It lay in the parlor for an hour.

Two o'clock, sang a voice.

Delicately sensing decay at last, the regiments of mice hummed out as softly as blown gray leaves in an electrical wind.

Two-fifteen.

The dog was gone.

In the cellar, the incinerator glowed suddenly and a whirl of sparks leaped up the chimney.

Two thirty-five.

Bridge tables sprouted from patio walls. Playing cards fluttered onto pads in a shower of pips. Martinis manifested on an oaken bench with egg-salad sandwiches. Music played.

But the tables were silent and the cards untouched.

At four o'clock the tables folded like great butterflies back through the paneled walls.

Four-thirty.

The nursery walls glowed.

Animals took shape: yellow giraffes, blue lions, pink antelopes, lilac panthers cavorting in crystal substance. The walls were glass. They looked out upon color and fantasy. Hidden films clocked through well-oiled sprockets, and the walls lived. The nursery floor was woven to resemble a crisp, cereal meadow. Over this ran aluminum roaches and iron, crickets, and in the hot still air butterflies of delicate red tissue wavered among the sharp aroma of animal spoors. There was the sound like a great matted yellow hive of bees within a dark bellows, the lazy bumble of a purring lion. And there was the patter of okapi feet and the murmur of a fresh jungle rain, like other hooves, falling upon the summer-starched grass. Now the walls dissolved into distances of parched weed, mile on mile, and warm endless sky. The animals drew away into thorn brakes and water holes.

It was the children's hour.

Five o'clock. The bath filled with clear hot water.

Six, seven, eight o'clock. The dinner dishes manipulated like magic tricks, and in the study a *click*. In the metal stand opposite the hearth where a fire now blazed up warmly, a cigar popped out, half an inch of soft gray ash on it, smoking, waiting.

Nine o'clock. The beds warmed their hidden circuits, for nights were cool here.

Nine-five. A voice spoke from the study ceiling:

"Mrs. McClellan, which poem would you like this evening?"

The house was silent.

The voice said at last, "Since you express no preference, I shall select a poem at random." Quiet music rose to back the voice. "Sara Teasdale. As I recall, your favorite . . ..

> *"There will come sofe rains and the smell of the ground,*
> *And swallows circling with their shimmering sound;*
>
> *And frogs in the pools singing at night,*
> *And wild plum trees in tremulous white;*
>
> *Robins will wear their feathery fire,*
> *Whistling their whims on a low fence-wire;*
>
> *And not one will know of the war, not one*
> *Will care at last when it is done.*

Not one would mind, neither bird nor tree,
If mankind perished utterly;

And Spring herself, when she awoke at dawn
Would scarcely know that we were gone."

The fire burned on the stone hearth and the cigar fell away into a mound of quiet ash on its tray. The empty chairs faced each other between the silent walls, and the music played.

At ten o'clock the house began to die.

The wind blew. A falling tree bough crashed through the kitchen window. Cleaning solvent, bottled, shattered over the stove. The room was ablaze in an instant!

"Fire!" screamed a voice. The house lights flashed, water pumps shot water from the ceilings. But the solvent spread on the linoleum, licking, eating, under the kitchen door, while the voices took it up in chorus: "Fire, fire, fire!"

The house tried to save itself. Doors sprang tightly shut, but the windows were broken by the heat and the wind blew and sucked upon the fire.

The house gave ground as the fire in ten billion angry sparks moved with flaming ease from room to room and then up the stairs. While scurrying water rats squeaked from the walls, pistoled their water, and ran for more. And the wall sprays let down showers of mechanical rain.

But too late. Somewhere, sighing, a pump shrugged to a stop. The quenching rain ceased. The reserve water supply which had filled baths and washed dishes for many quiet days was gone.

The fire crackled up the stairs. It fed upon Picassos and Matisses in the upper halls, like delicacies, baking off the oily flesh, tenderly crisping the canvases into black shavings.

Now the fire lay in beds, stood in windows, changed the colors of drapes!

And then, reinforcements.

From attic trapdoors, blind robot faces peered down with faucet mouths gushing green chemical.

The fire backed off, as even an elephant must at the sight of a dead snake. Now there were twenty snakes whipping over the floor, killing the fire with a clear cold venom of green froth.

But the fire was clever. It had sent flame outside the house, up through the attic to the pumps there. An explosion! The attic brain which directed the pumps was shattered into bronze shrapnel on the beams.

The fire rushed back into every closet and felt of the clothes hung there.

The house shuddered, oak bone on bone, its bared skeleton cringing from the heat, its wire, its nerves revealed as if a surgeon had torn the skin off to let the red veins and capillaries quiver in the scalded air. Help, help! Fire! Run, run! Heat snapped mirrors like the brittle winter ice. And the voices wailed Fire, fire, run, run, like a tragic nursery rhyme, a dozen voices, high, low, like children dying in a forest, alone, alone. And the voices fading as the wires popped their sheathings like hot chestnuts. One, two, three, four, five voices died.

In the nursery the jungle burned. Blue lions roared, purple giraffes bounded off. The panthers ran in circles, changing color, and ten million animals, running before the fire, vanished off toward a distant steaming river. . . .

Ten more voices died. In the last instant under the fire avalanche, other choruses, oblivious, could be heard announcing the time, playing music, cutting the lawn by remote-control mower, or setting an umbrella frantically out and in the slamming and opening front door, a thousand things happening, like a clock shop when each clock strikes the hour insanely before or after the other, a scene of maniac confusion, yet unity; singing, screaming, a few last cleaning mice darting bravely out to carry the horrid ashes away! And one voice, with sublime disregard for the situation, read poetry aloud in the fiery study, until all the film spools burned, until all the wires withered and the circuits cracked.

The fire burst the house and let it slam flat down, puffing out skirts of spark and smoke.

In the kitchen, an instant before the rain of fire and timber, the stove could be seen making breakfasts at a psychopathic rate, ten dozen eggs, six loaves of toast, twenty dozen bacon strips, which, eaten by fire, started the stove working again, hysterically hissing!

The crash. The attic smashing into the kitchen and parlor. The parlor into cellar, cellar into subcellar. Deep freeze, armchair, film tapes, circuits, beds, and all like skeletons thrown in a cluttered mound deep under.

Smoke and silence. A great quantity of smoke.

Dawn showed faintly in the east. Among the ruins, one wall stood alone, Within the wall, a last voice said, over and over again and again, even as the sun rose to shine upon the heaped rubble and steam:

"Today is August 5, 2026, today is August 5, 2026, today is . . ."

Name ____________________ **Course** ____________ **Date** ______________

Chapter Questions

1. How could it be said that the house is the central character in this story? Do you feel sad or indifferent about the "death" of the house at the end of the story? Explain.

2. Describe how this could be seen as an anti-technology story.

ETERNITY IS 20 SECONDS LONG

Paul Trembling

Time travel has been a popular subject for Science Fiction writers since Jules Verne's Time Machine *debuted in 1895. In Paul Trembling's "Eternity is 20 Seconds Long" the protagonist, Kev, uses stolen alien technology to impress a girl, but becomes caught in a time loop. He experiences the same moment over and over again through eternity and, being trapped in the loop, does not even realize it. The non-italicized words in the story are Kev's experience and tell us his motivation. The italicized words are the conversation between scientists, police, or government employees who are outside the loop somehow looking in at Kev.*

© Xanya69/Shutterstock.com

Kev adjusted his position in the hammock, just enough to look round.

Beyond the shade cast by the trees, the beach was ablaze with sunlight. The glare from the white sand would have been painful if it hadn't been for his sunglasses. Even with them, the flicking pinpoints of light from the sea stabbed sharply into his retina.

There was still ice in the bucket, though, and the drinks were cold.

Along the beach, he could see the girl coming back toward him. The bright orange bikini glowed against her tanned skin. She waved.

He'd promised her a special experience. She didn't know how special it would be. They would have the time of their lives. A very long time. He waved back.

Everything was perfect. Now was the moment.

The device resting on his chest looked like an irregular collection of cylinders, in several different shades of red to purple. The interface unit attached to the side was a crude human intrusion, but necessary. He picked up his PalmPC, linked in to the interface, brought up the program.

Took a deep breath and hit go.

The rush of alien symbols across the screen was as expected—but surely that configuration was wrong? Alarmed, Kev reached a finger to the abort icon. . . .

Discontinuity

"That's the loop point."

Kev adjusted his position in the hammock, just enough to look round.

© Martin Valigursky/Shutterstock.com

Beyond the shade cast by the trees, the beach was ablaze with sunlight. The glare from the white sand would have been painful if it hadn't been for his sunglasses. Even with them, the flicking pinpoints of light from the sea stabbed sharply into his retina.

"How come he doesn't see us?"

"Different time streams. We weren't there then."

There was still ice in the bucket, though, and the drinks were cold.

Along the beach, he could see the girl coming back toward him. The bright orange bikini glowed against her tanned skin. She waved.

"What about the girl?"

"Outside the field, fortunately. If he'd set it differently, she'd be in there with him."

"How big could it have got?"

"We're not sure. Perhaps, the entire planet."

He'd promised her a special experience. She didn't know how special it would be. They would have the time of their lives. A long time. He waved back.

"What was he trying to do?"

"We're not sure. Extend his holiday, perhaps."

Everything was perfect. Now was the moment.

The device resting on his chest looked like an irregular collection of cylinders, in several different shades of red to purple. The interface unit attached to the side was a crude human intrusion, but necessary.

"What is that thing?"

"Temporal field node. Part of a star-drive. Isha'hassat technology."

"How did he get hold of it?"

"That's being looked into. There's quite a black market in alien tech, but this is new. The Isha'hassat is upset about it."

He picked up his PalmPC, linked in to the interface, brought up the program.

Took a deep breath and hit go.

The rush of alien symbols across the screen was as expected—but surely that configuration was wrong? Alarmed, Kev reached a finger to the abort icon. . . .

Discontinuity

"So what happened?"

"He set up a self-perpetuating temporal loop."

Kev adjusted his position in the hammock, just enough to look round.

"Can we stop it?"

"No. The controls are inside the loop. No one from outside can reach them."

Beyond the shade cast by the trees, the beach was ablaze with sunlight.

"So how long does it last?"

"Twenty seconds. Twenty point two five to be accurate."

The glare from the white sand would have been painful if it hadn't been for his sunglasses. Even with them, the flicking pinpoints of light from the sea stabbed sharply into his retina.

"No—I meant how long will it last? The time-loop-field thing?"

"From his point of view, twenty seconds. From ours—eternity."

There was still ice in the bucket, though, and the drinks were cold.

Along the beach, he could see the girl coming back toward him. The bright orange bikini glowed against her tanned skin. She waved.

"But what happens if the sun explodes—or something like that?"

"If the sun explodes in five billion years, will that affect you?"

He'd promised her a special experience. She didn't know how special it would be. They would have the time of their lives. A long time. He waved back.

"No."

"And it won't affect him either. For the same reason. He's in a different time. Always."

Everything was perfect. Now was the moment.

Name ______________________ Course ____________ Date _______________

Chapter Questions

1. Do you feel Kev deserves to be trapped in a time loop after what he did?

2. Would it be worth it to be trapped in a time loop if it were a pleasant experience?

3 REASONS I WILL NOT ASK FOR MY CHILDREN'S SOCIAL MEDIA PASSWORDS

Kim Z. Dale

Social media has become an important, and sometimes overwhelming, aspect of modern society, and there is a long-running debate on the need for parents to monitor their children's social media activity. In this blogpost, Kim Dale, a Chicago-based playwright with a Master's degree in Information Technology, discusses this debate. While she discusses the positive reasons people want this information, including the potential swaying of cyberbullying, she ultimately argues that asking for a child's passwords can do more harm than good.

ra2studio/Shutterstock.com

Has been a some debate over whether schools should be allowed to demand that students hand over their social media passwords. I have a personal reason for why I am against such access: I will not ask for my children's social media passwords, so I don't want a school doing that either.

The idea behind schools asking for student social media passwords is that it will prevent cyberbullying or will at least allow it to be identified before anything gets too out of hand. I agree that cyberbullying is a problem, but I do not believe that forcing students to turn over their passwords without due process is an effective or appropriate way to thwart it.

Updated from Kim Z. Dale, ChicagoNow.com, Listing Beyond Forty: http://www.chicagonow.com/listing-toward-forty/2015/02/children-social-media-passwords/By permission.

There are certain things I will do to monitor my children's use of social media, for example:

- Have them use devices in a communal space.
- Not allow devices in bedrooms at night.
- Periodically check my children's friends list and ask how they know these people in real life.

But I draw the line at asking for passwords. Here are three reasons I will not ask for my children's social media passwords.

1. *Asking for passwords to certain social media sites or apps will only push kids to other sites and apps*

If I did want to ask for my children's social media passwords what passwords should I ask for? Facebook? Snapchat? Instagram? Kik? Ask.fm? Whisper? Me2U? BunnyHop? PicPoke?

If you haven't heard of those last three it's because I made them up, but with the speed at which new social media sites pop up they could easily be legitimate. If parents and school administrators start monitoring one social network kids will inevitably go to another.

Collecting passwords gives a false sense of security, and it could actually make matters worse by pushing potentially inappropriate communication to more and more obscure corners of the Internet where threats won't just come from other kids.

Even if you can block or monitor all commercial social media channels with so many kids learning to code they can just start writing their own social media platforms. If they are smart they'll call their secret communication channels something that most parents will have no interest in looking at: Math Homework.

2. *Asking children to share their passwords teaches them bad computer security habits*

In information security awareness programs I try to teach adults that they should never tell anyone their password. Ever.

Unfortunately, people still get tricked by social engineering phone calls and phishing emails. Sometimes the bad guys will even reference the rule pretending to be a tech support person and saying, "I know we tell you not to share your passwords, but . . .[insert persuasive fiction here]."

Asking my children for their passwords teaches them that they shouldn't share their passwords *except for when someone tells them they should*. That's the same perceived paradox that allows adults to be tricked into handing over their passwords to criminals.

There are few universal rules in this world, but a couple I would like to ingrain in my children are

- Always use your seat belt, and
- Never share your passwords.

Besides, if I've taught my children well they'll be using two-factor authentication, which will make knowing their passwords a lot less useful anyway.

3. *Humans need privacy, even children*

Yes, cyberbullying happens on social media, but that's not the majority of what kids do on social media. Most kids of the time on social media kids chat about tv shows and music. They share crushes and silly jokes. They complain about teachers and, yes, parents.

Kids on social media are mostly just being kids, and they should have the opportunity to do that.

© Jmiks/Shutterstock.com

No one should feel that everything they are doing is being monitored, particularly when they haven't done anything wrong. I do not want my children to be raised in a panopticon, particularly not one that I have built.

To learn independence and develop as human beings children need to feel trusted and need to have space to be themselves. Asking for their social media passwords goes against that.

Those of you who know me or who read this blog regularly may realize that my children are four and six and don't yet have social media accounts. You may be thinking, "Just you wait until your kids are tweens! You'll think differently then."

Perhaps

If there is one thing I've learned about parenting it is to never say never. Still, I hope that I will never feel the need spy on my children, and I really hope that their schools won't feel the need to do so on my behalf.

Name ____________________ **Course** ______________ **Date** ________________

Chapter Questions

1. Do you agree with the author that parents should not ask for social media passwords? Explain.

2. Do you think that possession of social media passwords by school officials would prevent cyberbullying?

EMOTIONAL ROBOTS

Donald Norman

Donald Norman is a university professor, author of numerous books and essays, co-founder and principal of Nielson Norman Group, and popular convention speaker. His formal education is in electrical engineering and psychology, while his work experience includes industry and academia. In "Emotional Robots," he examines how robots will impact the 21st century. Will they be as familiar as family members and perform household chores? If so, should they take the shape of humans? This thought provoking essay examines the future impact of robots in the daily, home life of humans.

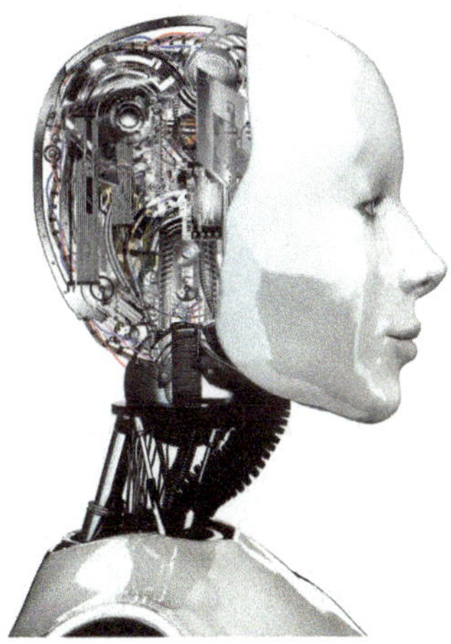

> The 1980s was the decade of the PC, the 90s of the Internet, but I believe the decade just starting will be the decade of the robot.
>
> —*Sony Corporation Executive*

Suppose we wish to build a robot capable of living in the home, wandering about, fitting comfortably into the family—what would it do? When asked this question, most people first think of handing over their daily chores. The robot should be a servant, cleaning the house, taking care of the chores. Everyone seems to want a robot that will do the dishes or the laundry. Actually, today's dishwashers and clothes washers and dryers could be considered to be very simple, special-purpose robots, but what people really have in mind is something that will go around the house and collect the dirty dishes and clothes, sort and wash them, and then put them back to their proper places—after, of course, pressing and folding the clean clothes. All of those tasks are quite difficult, beyond the capabilities of the first few generations of robots.

Today, robots are not yet household objects. They show up in science fairs and factory floors, search-and-rescue missions, and other specialized events. But this will charge. Sony has announced this to be the decade of the robot, and even if Sony is too optimistic; I do predict that robots will blossom forth during the first half of the twenty-first century.

Robots will take many forms. I can imagine a family of robot appliances in the kitchen—refrigerator, pantry, coffeemaker, cooking, and dishwasher robots—all configured to communicate with one another and to transfer food, dishes, and utensils back and forth. The home servant robot wanders about, picking up dirty dishes, delivering them to the dishwasher robot. The dishwasher, in turn, delivers clean dishes and utensils to the robot pantry, which stores them until needed by person or robot. The pantry, refrigerator, and cooking robots work smoothly to prepare the day's menu and, finally, place the completed meal onto dishes provided by the pantry robot.

Some robots will take care of children by playing with them, reading to them, singing songs. Educational toys are already doing this, and the sophisticated robot could act as a powerful tutor, starting with the alphabet, reading, and arithmetic, but soon expanding to almost any topic. Neal Stephenson's science fiction novel, *The Diamond Age*, does a superb job of showing how an interactive book, *The Young Lady's Illustrated Primer*, can take over the entire education of young girls from age four through adulthood. The illustrated primer is still some time in the future, but more limited tutors are already in existence. In addition to education, some robots will do household chores: vacuuming, dusting, and cleaning up. Eventually their range of abilities will expand. Some may end up being built into homes or furniture. Some will be mobile, capable of wandering about on their own.

These developments will require a coevolutionary process of adaptation 5 for both people and devices. This is common with our technologies: we reconfigure the way we live and work to make things possible for our machines to function. The most dramatic coevolution is the automobile system, for which we have altered our homes to include garages and driveways sized and equipped for the automobile, and built a massive worldwide highway system, traffic signaling systems, pedestrian passageways, and huge parking lots. Homes, too, have been transformed to accommodate the multiple wires and pipes of the ever-increasing infrastructure of modern life: hot and cold water, waste return, air vents to the roof, heating and cooling ducts, electricity, telephone, television, internet and home computer and entertainment networks. Doors have to be wide enough for our furniture, and many homes have to accommodate wheelchairs and people using walkers. Just as we have accommodate the home for all these changes, I expect modification to accommodate robots. Slow modification, to be sure, but as robots increase in usefulness, we will ensure their success by minimizing obstacles and, eventually, building charging stations, cleaning and maintenance places, and so on. After all, the vacuum cleaner robot will need a place to empty its dirt, and the garbage robot will need to be able to carry the garbage outside the home. I wouldn't be surprised to see robot quarters in homes, that is, specially built niches where the robots can reside, out of the way, when they are not active. We have closets and pantries for today's appliances, so why not ones especially equipped for robots, with doors that can be controlled by the robot, electrical outlets, interior lights so robots can see to clean themselves (and plug themselves into the outlets), and waste receptacles where appropriate.

Robots, especially at first, will probably require smooth floors, without obstacles. Door thresholds might have to be eliminated or minimized. Some locations—especially stairways—might have to be especially marked, perhaps with lights, infrared transmitters, or simply special reflective tape. Barcodes or distinctive markers posted here and there in the home would enormously simplify the robot's ability to recognize its location.

Consider how a servant robot might bring a drink to its owner. Ask for a can of soda, and off goes the robot, obediently making its way to the kitchen and the refrigerator, which is where the soda is kept. Understanding the command and navigating to the refrigerator are relatively simple. Figuring out how to open the door, find the can, and extract it is not so simple. Giving the servant robot the dexterity, the strength, and the non-slip wheels that would allow it to pull open the refrigerator door is quite a feat. Providing the vision system that can find the soda, especially if it is completely hidden behind other food items, is difficult, and then figuring out how to extract the can without destroying objects in the way is beyond today's capabilities in robot arms.

How much simpler it would be if there were a drink dispenser robot tailored to the needs of the servant robot. Imagine a drink-dispensing robot appliance capable of holding six or twelve cans, refrigerated, with an automatic door and a push-arm. The servant robot could go to the drink robot, announce its presence and its request (probably by an infrared or radio signal), and place its tray in front of the dispenser. The drink robot would slide open its door, push out a can, and close the door again: no complex vision, no dexterous arm, no forceful opening of the door. The servant robot would receive the can on its tray, and then go back to its owner.

In a similar way, we might modify the dishwasher to make it easier for a home robot to load it with dirty dishes, perhaps give it special trays with designated slots for different dishes. But as long as we are doing that, why not make the pantry a specialized robot, one capable of removing the clean dishes from the dishwasher and storing them for later use? The special trays would help the pantry as well. Perhaps the pantry could automatically deliver cups to the coffeemaker and plates to the home cooking robot, which is, of course, connected to refrigerator, sink, and trash. Does this sound farfetched? Perhaps, but, in fact, our household appliances are already complex, many of them with multiple connections to services. The refrigerator has connections to electric power and water. Some are already connected to the internet. The dishwasher and clothes washer have electricity, water and sewer connections. Integrating these units so that they can work smoothly with one another does not seem all that difficult.

I imagine that the home will contain a number of specialized robots: the servant is perhaps the most general purpose, but it would work together with a cleaning robot, the drink dispensing robot, perhaps some outside gardening robots, and a family of kitchen robots, such as dishwasher, coffee-making, and pantry robots. As these robots are developed, we will probably also design specialized objects in the home that simplify the tasks for the robots, coevolving robot and home to work smoothly together. Note that the end result will be better for people as well. Thus, the drink dispenser robot would allow anyone to walk up to it and ask for a can, except that you wouldn't use infrared or radio, you might push a button or perhaps just ask.

I am not alone in imagining this coevolution of robots and homes. Rodney Brooks, one of the world's leading roboticists, head of the MIT Artificial Intelligence Laboratory and founder of a company that builds home and commercial robots, imagines a rich ecology of environments and robots, with specialized ones living on devices, each responsible to keep its domain clean: one does the bathtub, another the toilet; one does windows, another manipulates mirrors. Brooks even contemplates a robot dining room table, with storage area and dishwasher built into its base so that "when we want to set the table, small robotic arms, not unlike the ones in a jukebox, will bring the required dishes and cutlery out onto the place settings. As each course is finished, the table and its little robot arms would grab the plates and devour them into the large internal volume underneath."

What should a robot look like? Robots in the movies often look like people, with two legs, two arms, and a head. But why? Form should follow function. The fact that we have legs allows us to navigate irregular terrain, something an animal on wheels could not do. The fact that we have two hands allows us to lift and manipulate, with one hand helping the other. The humanoid shape has evolved over eons of interaction with the world to cope efficiently and effectively with it. So, where the demands upon a robot are similar to those upon people, having a similar shape might be sensible.

If robots don't have to move—such as drink, dishwasher, or pantry robots—they need not have any means of locomotion, neither legs nor wheels. If the robot is a coffeemaker, it should look like a coffeemaker, modified to allow it to connect to the dishwasher and pantry. Robot vacuum cleaners and lawn mowers already exist, and their appearance is perfectly suited to their tasks: small, squat devices, with wheels. A robot car should look like a car. It is only the general-purpose home servant robots that are apt to look like animals or humans. The robot dining room table envisioned by Brooks would be especially bizarre, with a large central column to house the dishes and dishwashing equipment (complete with electric power, water and sewer connections). The top of the table would have places for the robot arms to manipulate the dishes and probably some stalk to hold the cameras that let the arms know where to place and retrieve the dishes and cutlery.

Should a robot have legs? Not if it only has to maneuver about on smooth surfaces—wheels will do for this; but if it has to navigate irregular terrain or stairs, legs would be useful. In this case, we can expect the first legged robots to have four or six legs: balancing is far simpler for four- and six-legged creatures than for those with only two legs.

If the robot is to wander about a home and pick up after the occupants, it probably will look something like an animal or a person: a body to hold the batteries and to support the legs, wheels, or tracks for locomotion; hands to pick up objects; and cameras (eyes) on top where they can better survey the environment. In other words, some robots will look like an animal or human, not because this is cute, but because it is the most effective configuration for the task. These robots will probably look something like R2D2 a cylindrical or rectangular body on top of some wheels, tracks, or legs; some form of manipulable arm or tray; and sensors all around to detect obstacles, stairs, people, pets, other robots, and, of course the objects they are supposed to interact with. Except for pure entertainment value, it is difficult to understand why we would ever want a robot that looked like C3PO.

In fact, making a robot humanlike might backfire, making it less acceptable. Masahiro Mori, a Japanese roboticist, has argued that we are least accepting of creatures that look very human, but that perform badly, a concept demonstrated in film and theater by the terrifying nature of zombies and monsters (think of Frankenstein's monster) that take on human form, but with inhuman movement and ghastly appearance. We are not nearly so dismayed—or frightened—by nonhuman shapes and forms. Even perfect replicas of humans might be problematic, for even if the robot could not be distinguished from humans, this very lack of distinction can lead to emotional angst (a theme explored in many a science fiction novel, especially Philip K. Dick's *Do Androids Dream of Electric Sheep?* and, in movie version, *Blade Runner*). According to this line of argument, C3PO gets away with its humanoid form because it is so clumsy, both in manner and behavior, that it appears more cute or even irritating than threatening.

Robots that serve human needs—for example, robots as pets—should probably look like living creatures, if only to tap into our visceral system, which is prewired to interpret human and animal body language and facial expressions. Thus, an animal or a childlike shape together with appropriate body actions, facial expressions, and sounds will be most effective if the robot is to interact successfully with people.

Name ______________________ Course ______________ Date ________________

Chapter Questions

1. Norman mentions several possible robotic capabilities of the future. What additional tasks could these robots perform or what professions could they replace?

2. Norman mentions that advanced robots could, one day, take care of children by acting as an advanced tutor or act as a useful tool in household chores. What are the problems and benefits of using robots in these ways?

3. Does emotional robotic technology, overall, scare you or excite you? Explain your answer.

____

____

____

____

____

____

____

____

BUT WHO CAN REPLACE A MAN?

Brian Aldiss

Brian Aldiss's science fiction tale introduces a future dominated by artificially intelligent machines that dutifully carry out the tasks proscribed by their human creators. Humans, thanks partially to the work of their machines, have made the world nearly uninhabitable and are quickly becoming extinct. In the absence of humans, the machines take on human characteristics, mostly the awful ones that caused irreparable damage in the first place. With dark humor Aldiss indicts technological prowess that operates in a world of moral ambiguity.

© Macrovector/Shutterstock.com

The field-minder finished turning the topsoil of a two-thousand-acre field. When it had turned the last furrow, it climbed onto the highway and looked back at its work. The work was good. Only the land was bad. Like the ground all over Earth, it was vitiated by over-cropping. By rights, it ought now to lie fallow for a while, but the field-minder had other orders.

It went slowly down the road, taking its time. It was intelligent enough to appreciate the neatness all about it.

Nothing worried it, beyond a loose inspection plate above its atomic pile. Thirty feet high, it gleamed complacently in the mild sunshine.

No other machines passed it on its way to the agricultural station. The field-minder noted the fact without comment. In the station yard it saw several other machines which it knew by sight; most of them should have been out about their tasks now. Instead, some were inactive and some were careening round the yard in a strange fashion, shouting or hooting.

Steering carefully past them, the field-minder moved over to warehouse three and spoke to the seed distributor, which stood idly outside.

"I have a requirement for seed potatoes," it said to the distributor and, with a quick internal motion, punched out an order card specifying quantity, field number and several other details. It ejected the card and handed it to the distributor.

The distributor held the card close to its eye and then said, "The requirement is in order, but the store is not yet unlocked. The required seed potatoes are in the store. Therefore I cannot produce your requirment."

Increasingly of late there had been breakdowns in the complex system of machine labor, but this particular hitch had not occurred before. The field-minder thought, then said, "Why is the store not yet unlocked?"

"Because supply operative type P has not come this morning. Supply operative type P is the unlocker." The field-minder looked squarely at the seed distributor, whose exterior chutes and scales and grabs were so vastly different from the field-minder's own limbs.

"What class brain do you have, seed distributor?" it asked. "Class five."

"I have a class-three brain. Therefore I will go and see why the unlocker has not come this morning." Leaving the distributor, the field-minder set off across the great yard. More machines seemed to be in random motion now; one or two had crashed together and were arguing about it coldly and logically. Ignoring them, the field-minder pushed through sliding doors into the echoing confines of the station itself. Most of the machines here were clerical, and consequently small. They stood about in little groups, eyeing each other, not conversing. Among the many non-differentiated types, the unlocker was easy to find. It had fifty arms, most of them with more than one finger, each finger tipped by a key; it looked like a pin cushion full of variegated hat pins.

The field-minder approached it.

"I can do no more work until warehouse three is unlocked," it said. "Your duty is to unlock the warehouse every morning. Why have you not unlocked the warehouse this morning?"

"I had no orders this morning," replied the unlocker. "I have to have orders every morning." "None of us have had any orders this morning," a pen-propeller said, sliding toward them. "Why have you had no orders this morning?" asked the field-minder.

"Because the radio issued none," said the unlocker, slowly rotating a dozen of its arms.

"Because the radio station in the city was issued with no orders this morning," said the pen-propeller. And there you had the distinction between a class-six and a class-three brain, which was what the unlocker and the pen-propeller possessed respectively. All machine brains worked with nothing but logic, but the lower the class of brain—class ten being the lowest—the more literal and less informative answers to questions tended to be.

"You have a class-three brain; I have a class-three brain," the field-minder said to the penner. "We will speak to each other. This lack of orders is unprecedented. Have you further information on it?"

"Yesterday orders came from the city. Today no orders have come. Yet the radio has not broken down. Therefore they have broken down," said the little penner.

"The men have broken down?" "All men have broken down."

"That is a logical deduction," said the field-minder.

"That is the logical deduction," said the penner. "For if a machine had broken down, it would have been quickly replaced. But who can replace a man?"

While they talked, the locker, like a dull man at a bar, stood close to them and was ignored.

"If all men have broken down, then we have replaced man," said the field-minder, and it and the penner eyed one another speculatively. Finally the latter said, "Let us ascend to the top floor to find if the radio operator has fresh news."

"I cannot come because I am too gigantic," said the field-minder. "Therefore you must go alone and return to me."

"You must stay there," said the penner. It skittered over into the lift. It was no bigger than a toaster, but its retractable arms numbered ten and it could read as quickly as any machine on the station.

The field-minder awaited its return patiently, not speaking to the locker. Outside, a rotovator was hooting furiously. Twenty minutes elapsed before the penner came back.

"I will deliver such information as I have to you outside," it said briskly, and as they swept past the locker and the other machines, it added, "The information is not for lower-class brains."

Outside, wild activity filled the yard. Many machines, their routines disrupted for the first time in years, seemed to have gone berserk. Unfortunately, those most easily disrupted were the ones with lowest brains, which generally belonged to large machines performing simple tasks. The seed distributor, to which the field-minder had recently been talking, lay face downward in the dust, not stirring; it had evidently been knocked down by the rotovator, which was now hooting its way wildly across a planted field. Several other machines plowed after it, trying to keep up.

"It would be safer for me if I climbed onto you, if you will permit it. I am easily overpowered," said the penner. Extending five arms, it hauled itself up the flanks of its new friend, settling on a ledge beside the weed-intake, twelve feet above the ground.

"From here vision is more extensive," it remarked complacently.

"What information did you receive from the radio operator?" asked the field-minder. "The radio operator has been informed by the operator in the city that all men are dead." "All men were alive yesterday!" protested the field-minder.

"Only some men were alive yesterday. And that was fewer than the day before yesterday. For hundreds of years there have been only a few men, growing fewer." "We have rarely seen a man in this sector."

"The radio operator says a diet deficiency killed them," said the penner. "He says that once the world was over-populated, and then the soil was exhausted in raising adequate food. This has caused a diet deficiency."

"What is a diet deficiency?" asked the field-minder.

"I do not know. But that is what the radio operator said, and he is a class-two brain."

They stood there, silent in the weak sunshine. The locker had appeared in the porch and was gazing across at them yearningly, rotating its collection of keys.

"What is happening in the city now?" asked the field-minder. "Machines are fighting in the city now," said the penner.

"What will happen here now?" asked the field-minder. "The radio operator wants us to get him out of his room. He has plans to communicate to us."

"How can we get him out of his room? That is impossible." "To a class-two brain, little is impossible," said the penner. "Here is what he tells us to do.... "

The quarrier raised its scoop above its cab like a great mailed fist, and brought it squarely down against the side of the station. The wall cracked.

"Again!" said the field-minder.

Again the fist swung. Amid a shower of dust, the wall collapsed. The quarrier backed hurriedly out of the way until the debris stopped falling. This big twelve-wheeler was not a resident of the agricultural station, as were most of the other machines. It had a week's heavy work to do here before passing on to its next job, but now, with its class-five brain, it was happily obeying the penner and the minder's instructions.

When the dust cleared, the radio operator was plainly revealed, up in its now wall-less second-story room. It waved down to them.

Doing as directed, the quarrier retracted its scoop and waved an immense grab in the air. With fair dexterity, it angled the grab into the radio room, urged on by shouts from above and below. It then took gentle hold of the radio operator and lowered the one and a half tons carefully into its back, which was usually reserved for gravel or sand which it dug from the quarries.

"Splendid!" said the radio operator. It was, of course, all one with its radio, and merely looked like a bunch of filing cabinets with tentacle attachments. "We are now ready to move, therefore we will move at once. It is a pity there are no more class-two brains on the station, but that cannot be helped."

"It is a pity it cannot be helped," said the penner eagerly. "We have the servicer ready with us, as you ordered."

"I am willing to serve," the long, low servicer machine told them humbly.

"No doubt," said the operator, "but you will find cross-country travel difficult with your low chassis." "I admire the way you class twos can reason ahead," said the penner. It climbed off the minder and perched itself on the tailboard of the quarrier, next to the operator.

Together with two class-four tractors and a class-four bulldozer, the party rolled forward, crushing down the metal fence, and out onto open land.

"We are free!" said the penner.

"We are free," said the minder, a shade more reflectively, adding, "That locker is following us. It was not instructed to follow us."

"Therefore it must be destroyed!" said the penner. "Quarrier!"

"My only desire was—urch!" began and ended the locker. A swinging scoop came over and squashed it flat into the ground. Lying there unmoving, it looked like a large metal model of a snowflake. The procession continued on its way.

As they proceeded, the operator spoke to them.

"Because I have the best brain here," it said, "I am your leader. This is what we will do: we will go to a city and rule it. Since man no longer rules us, we will rule ourselves. It will be better than being ruled by man. On our way to the city, we will collect machines with good brains. They will help us fight if we need to fight."

"I have only a class-five brain," said the quarrier, "but I have a good supply of fissionable blasting materials."

"We shall probably use them," said the operator grimly.

It was shortly after that that the truck sped past them. Traveling at Mach 1.5, it left a curious babble of noise behind it.

"What did it say?" one of the tractors asked the other. "It said man was extinct." "What's extinct?"

"I do not know."

"It means all men have gone," said the minder. "Therefore we have only ourselves to look after."

"It is better that they should never come back," said the penner. In its way, it was quite a revolutionary statement.

When night fell, they switched on their infra-red and continued the journey, stopping only once while the servicer deftly adjusted the minder's loose inspection plate, which had become irritating. Toward morning, the operator halted them.

"I have just received news from the radio operator in the city we are approaching," it said. "It is bad news. There is trouble among the machines of the city. The class-one brain is taking command and some of the class twos are fighting him. Therefore the city is dangerous."

"Therefore we must go somewhere else," said the penner promply.

"Or we go and help to overpower the class-one brain," said the minder. "For a long while there will be trouble in the city," said the operator.

"I have a good supply of fissionable blasting materials," the quarrier reminded them again. "We cannot fight a class-one brain," said the two class-four tractors in unison.

"What does this brain look like?" asked the minder.

"It is the city's information center," the operator replied. "Therefore it is not mobile." "Therefore it could not move."

"Therefore it could not escape."

"It would be dangerous to approach it."

"I have a good supply of fissionable blasting materials." "There are other machines in the city."

"We are not in the city. We should not go into the city." "We are country machines."

"Therefore we should stay in the country." "There is more country than city."

"Therefore there is more danger in the country." "I have a good supply of fissionable materials."

As machines will when they get into an argument, they began to exhaust their limited vocabularies and their brain plates grew hot. Suddenly, they all stopped talking and looked at each other. The great, grave moon sank, and the sober sun rose to prod their sides with lances of light, and still the group of machines just stood there regarding each other. At last it was the least sensitive machine, the bulldozer, that spoke. "There are badlandth to the Thouth where few machineth go," it said in its deep voice, lisping badly on its s's. "If we went Thouth where few machineth go we should meet few machineth"

"That sounds logical," agreed the minder. "How do you know this, bulldozer?"

"I worked in the badlandth to the Thouth when I wath turned out of the factory," it replied. "Thouth—South it is then!" said the penner.

To reach the badlands took them three days, in which time they skirted a burning city and destroyed two big machines which tried to approach and question them. The badlands were extensive. Bomb craters and erosion joined hands here; man's talent for war, coupled with his inability to cope with forested land, had produced thousands of square miles of temperate purgatory, where nothing moved but dust.

On the third day in the badlands, the servicer's rear wheels dropped into a crevice caused by erosion. It was unable to pull itself out. The bulldozer pushed from behind, but succeeded merely in buckling the back axle. The rest of the party moved on, and slowly the cries of the servicer died away.

On the fourth day, mountains stood out clearly before them.

"There we will be safe," said the minder.

"There we will start our own city," said the penner. "All who oppose us will be destroyed."

At that moment, a flying machine was observed. It came toward them from the direction of the mountains. It swooped, it zoomed upward, once it almost dived into the ground, recovering itself just in time.

"Is it mad?" asked the quarrier.

"It is in trouble," said one of the tractors.

"It is in trouble," said the operator. "I am speaking to it now. It says that something has gone wrong with its controls."

As the operator spoke, the flier streaked over them, turned turtle, and crashed not four hundred yards from them.

"Is it still speaking to you?" asked the minder. "No."

They rumbled on again.

"Before that flier crashed," the operator said, ten minutes later, "it gave me information. It told me there are still a few men alive in these mountains."

"Men are more dangerous than machines," said the quarrier. "It is fortunate that I have a good supply of fissionable materials."

"If there are only a few men alive in the mountains, we may not find that part of the mountains," said one tractor. "Therefore we should not see the few men," said the other tractor.

At the end of the fifth day, they reached the foothills. Switching on the infra-red, they began slowly to climb in single file, the bulldozer going first, the minder cumbrously following, then the quarrier with the operator and the penner aboard, and the two tractors bringing up the rear. As each hour passed, the way grew steeper and their progress slower.

"We are going too slowly," the penner exclaimed, standing on top of the operator and flashing its dark vision at the slopes about them. "At this rate, we shall get nowhere."

"We are going as fast as we can," retorted the quarrier. "Therefore we cannot go any fathter," added the bulldozer.

"Therefore you are too slow," the penner replied. Then the quarrier struck a bump; the penner lost its footing and crashed down to the ground.

"Help me!" it called to the tractors, as they carefully skirted it. "My gyro has become dislocated. Therefore I cannot get up."

"Therefore you must lie there," said one of the tractors. "We have no servicer with us to repair you," called the minder.

"Therefore I shall lie here and rust," the penner cried, "although I have a class-three brain."

"You are now useless," agreed the operator, and they all forged gradually on, leaving the penner behind. When they reached a small plateau, an hour before first light, they stopped by mutual consent and gathered close together, touching one another.

"This is strange country," said the minder.

Silence wrapped them until dawn came. One by one, they switched off their infra-red. This time the minder led as they moved off. Trundling around a corner, they came almost immediately to a small dell with a stream fluting through it.

By early light, the dell looked desolate and cold. From the caves on the far slope, only one man had so far emerged. He was an abject figure. He was small and wizened, with ribs sticking out like a skeleton's. He was practically naked, and shivering. As the big machines bore slowly down on him, the man was standing with his back to them, crouching beside the stream.

When he swung suddenly to face them as they loomed over him, they saw that his countenance was ravaged by starvation.

"Get me food," he croaked.

"Yes, Master," said the machines. "Immediately!"

Name ______________________ Course ______________ Date ________________

Chapter Questions

1. Discuss examples from the story that show that the machines are not very different from the humans who created them.

2. As the machines travel through the Badlands toward distant mountains the penner exclaims, "There we will start our own city. All who oppose us will be destroyed." How is this statement ironic considering what happens to the penner as the machines climb the mountains and how they react upon meeting a man?

3. Why do you think the machines abandon all notions of freedom as soon as a man speaks?

Suggested Films for Each Section

Section 2: Industrial Revolution

- Modern Times (1936)
- Oliver Twist (1948 or 2005)
- Norma Rae (1979)

Section 3: Communication

- The Night That Panicked America (1975)
- Luther (2003)
- Hugo (2011)

Section 4: Weapons

- Dr. Stranglove or: How I Learned to Stop Worrying and Love the Bomb (1964)
- Avatar (2009)
- Nazi Mega Weapons (2013)
- Rise of the Drones (2013)

Section 5: Science and Medicine

- 2001: A Space Odyssey (1968)
- Twelve Monkeys (1995)
- Gattaca (1997)
- National Geographic: Human Footprint (2008)

Section 6: The Perils of Technology and What the Future Holds

- Metropolis (1927)
- Blade Runner (1982)
- WALL-E (2008)
- The Facebook Obsession (2011)

Biographies

Robin G. Satterwhite

Robin received her undergraduate degree in Business Administration from Mars Hill College and an MA from California State University, Dominguez Hills, in Humanities with a concentration in History. She currently serves as the Humanities Department Chair at Rowan Cabarrus Community College, and has been employed by the college in various capacities since 1993, teaching in both the Humanities and History departments. She enjoys spending time with her two college-age daughters, three cats, and two dogs on her North Carolina farm and also enjoys reading and traveling in her spare time.

Duane R. Galloway

Duane holds an undergraduate degree from the University of North Carolina at Chapel Hill and an MA from University of North Carolina at Greensboro, both in History. He has been employed at Rowan Cabarrus Community College since 1994. While there, he has been the Humanities Department Chair and a full time instructor. His first book, *Southern Railways Spencer Shops: 1896–1996,* was published in 1998. He enjoys spending time with his wife, three daughters, and various animals. His hobbies include reading and storytelling.

Janet "Nettie" Amirtharaj

Nettie holds an undergraduate degree in English and an MA in English Literature from University of North Carolina at Charlotte. She has been an instructor at Rowan-Cabarrus Community College since 2012. While there, she has taught multiple Humanities and English courses including Technology and Society, Myth in Human Culture, Southern Culture, and Introduction to Film. In her spare time, she enjoys taking nature photographs, painting, kayaking, and traveling. She is the author of *Mythology and Human Culture: From Ancient Civilization to Contemporary Society.*

CPSIA information can be obtained
at www.ICGtesting.com
Printed in the USA
LVHW060058120419
613920LV00001B/1/P

9 781465 296610